Gender and Boyle's Law of Gases

Race, Gender, and Science

Anne Fausto-Sterling, *General Editor*

Elizabeth Potter

Gender and Boyle's Law of Gases

Indiana University Press

Bloomington and Indianapolis

This book is a publication of

Indiana University Press
601 North Morton Street
Bloomington, IN 47404-3797 USA

http://iupress.indiana.edu

Telephone orders 800-842-6796
Fax orders 812-855-7931
Orders by e-mail iuporder@indiana.edu

The paper used in this publication meets the minimum requirements
of American National Standard for Information Sciences—Permanence
of Paper for Printed Library Materials, ANSI Z39.48-1984.

Manufactured in the United States of America

Library of Congress Cataloging-in-Publication Data

Potter, Elizabeth.
     Gender and Boyle's law of gases / Elizabeth Potter.
          p.   cm. — (Race, gender, and science)
     Includes index.
          ISBN 0-253-33916-2 (cl : alk. paper) — ISBN 0-253-21455-6
(pa : alk. paper)
          1. Boyle's law. 2. Hylozoism. 3. Science—England—History—
17th century. I. Title. II.   Series.

QC161 .P68 2001
509.42'09'032—dc21

                                                                00-061398

     1 2 3 4 5   06 05 04 03 02 01

Contents

17.   Methodological Considerations   155

18.   "The Data Alone Proved Boyle's Hypothesis"   161

19.   Good Science   172

        Conclusion   180

        Notes   187

        Index   203

Acknowledgments

Audiences at colleges and universities all over the country were crucial to the making of this book, particularly those at Brown University, Duke University, Amherst College, Hamilton College, Haverford College, Stanford University, Swarthmore College, the University of California at Davis, and the University of California at San Francisco. I also benefited from the responses of fellow philosophers at many meetings of the Society for Women in Philosophy.

Special thanks are due to Linda Alcoff, Helen Longino, Lynn Nelson, Anne Fausto-Sterling, Nancy Tuana, and Alison Wylie, and to my great friend Dr. Fernando Rodriguez Casas. Both Mills College and Hamilton College granted me leaves to work on this book, for which I thank them.

I also thank the president and board of the Royal Society of London for permission to quote from the Boyle Papers and the Department of Special Collections, Stanford University Libraries, for permission to reproduce plates from *The Works of the Honourable Robert Boyle* (1772 edition).

Portions of chapter 19 first appeared in "Modeling the Gender Politics in Science," published in *Hypatia,* and in "Good Science and Good Philosophy of Science," published in *Synthese.* I thank the editors of these journals for permission to use this material.

Introduction

Feminist scholars of science have worked to recover the lost contributions of women to science, to examine scientific theories of women's nature, and to study the ways gender considerations have affected both the practice and the technical content of scientific theories, norms, and methods.[1] One of the greatest challenges has been to show the influence of gender considerations on the technical content of the physical sciences. To dash cold water on the feminist hunch that no area of science is immune to gender politics, doubters often ask, "What could gender have to do with something like Boyle's Law?" Boyle's Law of Gases, classically expressed as $k = pvt$ or $p_1v_1t_1 = p_2v_2t_2$, is supposed to provide an especially tough and resistant example because it is the mathematical expression of the scientifically established fact that (roughly speaking) the pressure, volume, and temperature of a gas are proportional to one another. The law says, for example, that if the volume of a gas remains the same, but the pressure increases, then the temperature of the gas will rise; if the volume increases and the temperature remains stable, the pressure will decrease.

Work for this book was undertaken following just such a challenge issued to Evelyn Fox Keller following a public presentation of the work for which she is now so well known. In the early 1980s, the Boston University Philosophy of Science Colloquium featured a two-day session on the emerging field of Gender and Science. Keller's lecture was followed by a very peppy, occasionally heated question period in the course of which one person, finding his objection answered not only elegantly but rather thoroughly, changed course and ended with, "Yeah, but you'll never show that gender affects something like Boyle's Law!" This book is a response to that challenge.

The first thing to note in responding is that Boyle's Law is not isolated from its scientific context; it follows from certain important assumptions and so belongs with a huge set of hypotheses and theories (some people refer to such a large set as a "paradigm"). To understand how Boyle arrived at the Ideal Gas Law, as it is also called, we must understand how he came to accept the set of hypotheses and theories providing the context for it. Fortunately, Boyle tells us a lot about the considerations that weighed in favor of that set of hypotheses and theories. That set is now referred to as "mechanism" or "the mechanistic paradigm." Usually, we will refer to the relevant set of theories and hypotheses simply as "mechanism" or, as Boyle sometimes referred to it, "the Mechanical or Corpuscular Philosophy."

Perhaps the principle of inertia best typifies mechanism; in Boyle's time, René Descartes expressed this fundamental principle by stating that matter remains at rest or in uniform motion in a straight line unless acted upon by some other body already in motion. Nowadays we would say "unless it is acted upon by some external force," but at that time forces were deemed "occult." It is hard for us to imagine what it was like when Boyle was working on such issues, for that was a time when many people held a fundamental idea precisely opposed to the principle of inertia, viz., that matter could move itself. These people were hylozoists. There were many different schools of hylozoism. Some hylozoists were animists and thought that all matter is alive and has consciousness. That is, humans have rational as well as perceptual faculties, but even rocks have some sort of low-level perception. Many animists believed that nature is alive and conscious in the way that a person is; some referred to "her" as Nature and some as the World Soul or World Spirit. Other hylozoists were not clearly animists, but still believed that matter could move itself. In fact, even mechanists were not always thorough-going mechanists. It is not at all clear, for example, that Boyle was—some scholars think that he maintained a residual belief in the activity, though not the animation, of some kinds of matter. See chapter 10 below.

To show how ideas about gender intersected Boyle's ideas about hylozoism and mechanism, we will first show that Robert Boyle thought about gender. In Part I, we will examine a few of the many pieces he wrote about women and about what kind of man can best do science. There we will see that his concept of the masculinity appropriate for the man of science had a silent opposite, a certain conception of femininity. According to these conceptions, manly men are well suited for science, but womanly women are not. Most Boyle scholars have not paid attention to his ideas about gender because gender has seemed irrelevant to his scientific work, particularly the work in chemistry for which he is famous. Gender disappears from standard accounts of how Boyle came to pursue a mechanistic research program leading him to the Gas Law; these accounts give a history of some of the ideas influencing his work but leave out others, such as class and gender, that also played a part in it. Historians and philosophers have assumed that standard accounts are historically adequate and that, in any case, Boyle's work wouldn't be good science if considerations of this sort influenced it. (We will return to this point below.) Part I ends with a standard account of Boyle's work leading to the Gas Law so that we can begin with an understanding of why it has been so hard for scholars to imagine what connection gender and class could have with Boyle's Law.

Our counter-argument to standard accounts is found in Part II. If we look at the social and intellectual context of his work, we find that Boyle was studying the writings of natural philosophers (our term "scientist" came along much later) working in all of the major paradigms in play at that time. But we also find that Boyle lived in England during one of the most turbulent periods in the history of that country, the English Civil War and Revolution (roughly, 1640–1661). The turbulence of this twenty-year period allowed the expression, orally and in print, of radical new ideas about social, political, economic, and gender equality. In Part II, we will examine many of these ideas and the activities of many of the women and men who advanced them. Historians of seventeenth-century science and religion have already shown that these radicals' ideas about equality found a basis in the hylozoic paradigm. The hylozoic understanding of the natural world had implications for their understanding of the social world—both the world they lived in and the world they aspired to live in. Others, however, even those who wanted to curtail the excesses of the Stuart kings and to have a stronger Parliament, were appalled by these radical egalitarian ideas. We will set out a fresh case for the thesis, put forward by some seventeenth-century historians, that Boyle considered the religious and social meaning of hylozoism in his decision to pursue mechanism. (Actually, as we will see, religious and social issues were inseparable at that time.) This book argues that while evidence, experimental and observational, provided a strong consideration in Boyle's choice of natural philosophy, he had recourse to other considerations as well because both mechanism and hylozoism had reasonable empirical adequacy, i.e., both accounted for the data reasonably well. Specifically, the implications of animism were a factor in Boyle's decision to reject it in all its forms—from relatively sober Aristotelianism to the lively animism of Helmont and other Paracelsians. And these implications included not only political and economic arrangements, but gender arrangements as well.

To see that both hylozoism and mechanism had reasonable degrees of empirical adequacy, we will examine the controversy between Boyle and Franciscus Linus. Linus gave a hylozoic interpretation of the experimental evidence Boyle offered in defense of the corpuscular philosophy. Attention to this particular controversy will allow us to see how Boyle produced his law as part of his work to defeat animism and defend mechanism. Boyle tells us about one of the experiments he conducted to support his hypothesis that the air has weight and spring (later referred to as "pressure") and to defeat the alternative hylozoic hypothesis that Nature abhors a vacuum and funiculi form to prevent one. He noticed that the experiment produced a series of observations indicating that the volume of the air is inversely proportional to its weight and spring. This is essentially Boyle's Law. But we will also see that there could have been a hylozoic alternative to Boyle's Law.

Most people assume that if beliefs about politics, economics, gender, religion, and so on influence scientific work, the result will be bad science. These beliefs are referred to as biases and the scientist is said to be biased. I do not agree with most people on this issue; therefore, at the end of Part II, I will show how it is possible for a background assumption about political or social arrangements to influence scientific work and the results still to be good science, with reasonable empirical adequacy and other scientific virtues. I maintain, therefore, that Robert Boyle did very good experimental work and deserves his role as a paradigm of the early modern scientist. I will use a model of scientific theories to show that when scientists are faced with competing hypotheses, both of which cover the available data, their decision in favor of one or the other is constrained by background assumptions. Anything can serve as a background assumption and scholars of science must look at each case to see what the background assumptions were. Thus I will argue, against scholars who present

Boyle's social and political assumptions as determining his scientific views, that Boyle's first concern was for his hypotheses to "accord well" with the data. But I will also argue, against scholars who have assumed or argued that Boyle's decision to take up the corpuscular philosophy was determined solely by the data, that background assumptions, including those about gender and class, influenced his choice of research program. His choice was constrained both by the data and by social and political assumptions.

Part One:
The Intersection
of Gender and Science

Now We See It. Now We Don't.

1
Now We See It

In this chapter, we will see that Robert Boyle thought about gender; chapter 8 below includes detailed discussion of his writings about femininity. Here we will see that as he thought and wrote about proper experimental method and other important aspects of the new experimental science, he pictured women in a complex relationship to the scientist and to scientific work.

Speaking to men about women, Robert Boyle once said, "I will not here examine, whether the ignorance wont to be imputed to women be their fault, or that of their accusers; and whether it is any natural want of capacity, or rather want of instruction, that keeps most of them from knowledge, though this regards not sexes. [Nor will I inquire] whether it be not our interest, or our envy, that makes women what we are wont to decry them for being. . . . " He is convinced that "to attain a great proficiency in knowledge," it is not necessary to be "so much as a man."[1]

Here, although Boyle refuses to consider whether women are born without the capacity for knowledge, he does not assume that they are; instead, he hints that women are made to be "what we are wont to decry them for being." If women simply lack instruction, they could be instructed; perhaps they could be instructed in natural philosophy. If they have the capacity for knowledge, perhaps they could participate in scientific work. However, Boyle never pursued these possibilities; in fact, as we examine his work to produce the new experimental science, we will find that he worked to produce a new form of masculinity conducive to the new science as he envisioned it, and that he did so by reinforcing a traditional form of femininity, one necessary for the new man of science. We will then see that modern experimental science was intensely gendered at its inception and that Boyle's gender ideology was in turn shaped by the demands of experimental science. In particular we will see that Boyle's accomplishments included not only the development of the experimental method and the lab report, but also, and through many of the same discursive maneuvers, the development of a man and woman appropriate to the new experimental science.

Boyle was very concerned about what it meant to be a woman and what it meant to be a man, but a strong challenge faces the claim that these concerns shaped and were shaped by what it meant to Boyle to be an experimental scientist or, to use his term, a "virtuoso." His work on gender appears to be merely interleaved with his work on experimental science; that is, he wrote about gender and he wrote about science, but because he did not explicitly write about gender and science, it is very difficult to see that these parts of his work are not only interleaved but mutually informing.

Although from a twentieth-century point of view we do not find a crisp distinction between Boyle's scientific and non-scientific works (both are usually addressed to friends and often employ the first person), he is nevertheless esteemed as the father of the lab report, the (relatively) straightforward description of an experiment and the conclusions based upon it. He is credited with being among the first to distinguish between scientific and non-scientific genres, as an integral part of his construction of experimental science. But from our point of view, it is unfortunate that Boyle created the distinction, for it allowed him to separate matters into the scientific and the personal or social, to create the idea of a chasm between them, and it suppresses the interaction between them. We must set his "non-scientific" works side by side with his "scientific" ones in order to see that Boyle helped to construct a new science which required that facts be produced through experiments properly conducted and attested, and that the new man of science be a chaste, modest heterosexual who desires yet eschews a sexually dangerous yet chaste and modest woman.

Although he wrote about fictional women, and revealed a rather more puritan than cavalier view of them, his relations with real women are more elusive. We do know that he never married. This was a bit unusual for men of his class; he was the youngest son of the infamous, fabulously wealthy Earl of Cork, and although the old Earl lost his life and most of his wealth in 1642 at the end of an Irish rebellion, his sons regained the estates and most of the wealth. Unlike many younger sons of the aristocracy, then, Robert Boyle had the money to marry. And letters to him and about him tell us that his friends were always trying to get him to marry somebody or other.[2]

But his own letters and other writings make it clear that he most definitely did not want to marry. For one thing, his older sister, Lady Katherine Ranelagh, served him in many of the ways a wife would have. When he first came to London in 1642, he stayed with her and she introduced him to her many important friends of the Parliamentary party (friends who helped him recover his Irish estates). Before he moved to Oxford in 1656, she stayed in his apartments there and worried that the rooms were too drafty for the hypochondriacal Boyle.[3] He lived the last twenty years of his life at her house on Pall Mall. But his sentiments, as we shall see, were not those of a man who favored marriage for anybody.

Though many Renaissance and early modern writers held, with medieval ones, that women should be chaste, silent, and obedient, Boyle explicitly discussed only their chastity and modesty. To Boyle, chastity was the most important characteristic women could have, and it was imperative for women because it was so difficult for them. He argued that women wanted to be whores. In a letter addressed simply to "Madam," Boyle remarked of women,

> I am confident that thousands would be whores could they but be so without being thought so and would not scruple at the acts of lust were not the imputation infamous and they more concerned in their reputations than in their consciences.[4]

His epistolary writings about women almost invariably picture them as seething with lust, their uncontrolled sexuality ready to explode. In a letter addressed to Mrs. Dury, he inveighs against the fictitious Corisca's using make-up and black patches and wearing dresses that reveal her bosom. Her naked breasts, he remarks, are like the famous northern Etna that "has not more snow at the top than fire underneath. One that were satirically given might say that she wears her breasts open to give cool vent to those lusts that burn within them" (BP, XXXII).

Why was Boyle so determined that women be chaste? The argument he took up most often against painting, baring the breasts, and other untoward behavior reveals that women must be chaste to keep men chaste. It turns out that women's chastity benefits men; it aids men's chastity. Women ought not wear make-up, for instance, because

> few men will be found refractory to believe that . . . these
> ladies desire this complexion to please others than their hus-
> bands; and that they are ambitious of Beauty for no very noble
> ends, that blush not to seek for it by such unhandsome means,
> not to mention, how much the suspicion of this truth invites
> loose gallants to tempt them that are suspected; and that such
> a false Beauty (when first supposed not to be so) does often-
> times occasion real sins by kindling desire whose guilt is not so
> barely imaginary as their causes.

This passage begins with concern that when men suspect that women wear make-up, the loose gallants among them will tempt those women to unchastity, but moves to concern that women wearing make-up often cause real sins by kindling guilty desire in men. Arguments against the naked-breasted Corisca move in the same way; Corisca should consider "what wanton flames this snow of her breasts kindles in those of others . . . whose concupiscences though she be not guilty of satisfying, yet she is of provoking. . . . [B]y this lure of a bare neck" she "exposes her chastity to unnecessary dangers," usurps "the devil's office," and "tempt[s] others to tempt her." In exposing her breasts, a woman endangers her own chastity, but it is the devil's work to tempt men (thereby endangering their chastity) to tempt her.

No wonder, then, that Boyle prefers a pure, iron lady: "Though I must tell you with all," he remarks toward the end of this letter, "that were my own fancy to be my chooser it would cry, give me a woman whose chastity is not only so constantly impregnable that no man can force it, but so notedly severe that no man dares besiege it." Such a woman would, indeed, be good insurance for male chastity! The re-markable desire expressed in this cry for a woman who is a fortress, impregnable to male force, daunting even attempts to take her, im-mediately raises the question why male chastity was so important for Boyle. The answer is found in certain of his non-scientific texts, where we find a close link between gender and science: more specifically, between chastity, undistracted devotion to God, and experimental science.

The first of these texts is *Some Motives and Incentives to the Love of God* or *Seraphick Love,* completed in 1648 and published in 1659,[5] a dis-course written as a letter to a friend called "Lindamor" (a name Boyle sometimes used as a pseudonym for his brother, Roger, Baron Brog-hill). Lindamor has been jilted by Hermione and Boyle writes to per-suade him to "transfigure" his love—in modern parlance, to subli-mate it—by turning it to God. Boyle remarks, "I can scarce disallow the being moderately in love, without being injurious to marriage, which is a relation, that though I can with much less reluctancy per-mit others than contract myself; yet dare I not absolutely condemn a condition of life as expedient to no man, without which even Paradise and innocence were not sufficient to complete the happiness of the first man" (*W*, I, 249).

This is an extraordinary remark. He would disallow even being moderately in love with a woman, except that disallowing love in-jures marriage, a contract that he is quite reluctant to enter. In fact, he would condemn marriage absolutely, but he dares not since God ordained it for Adam's happiness. He cites with approval St. Paul's argument for celibacy; the apostle tells us, he says, "that there is dif-ference betwixt married and single persons, the affections of the one being at liberty to devote themselves more undistractedly to God, whereas those of the other are distracted; (as Adam's were betwixt his Maker and his rib)" (*W*, I, 254).[6] Boyle encourages Lindamor to heed St. Paul's argument.

Passages such as this one suggest that Boyle's personal decision not to marry was determined, at least in part, by his professed desire to serve God. Birch reports that while he was in Geneva as a very young man, Boyle had undergone a conversion experience and had conse-crated his life to piety. And as James Jacob points out, "To [Boyle's] mind a consecration to piety was inconsistent with sensual gratificat-ion—and in a way that is not at all clear. He never married. Perhaps he saw marriage as being at odds, too, with the pursuit of piety" (Ja-cob, *Robert Boyle*, 39). From his own writings, Boyle appears to be an exception to the generalization he draws that most men will not turn from women to God until they have been jilted; "such is our frailty," he notes in *Seraphick Love*, "that . . . the amorous soul needs the harsh usage of a disdainful mistress to disgust it with its thraldom, and make it aspire to its more genuine and satisfying object" (*W*, I, 252). He himself did not need jilting. "Though his boiling youth did often very earnestly solicit to be employed in those culpable delights that are useful in and seem so proper for that season . . . ; yet did its im-

portunities meet ever with denials" (*W*, I, xxii). He said again and
again that *he* was never overcome by Cupid.[7] By his own criteria,
then, Boyle approached perfection, as John Evelyn pointed out. In a
letter to Boyle defending marriage against the attack on it in *Seraphick
Love,* Evelyn concluded that though Boyle was right, those of us who
are married must serve and love God "as well as we may in the condition we are assigned; which if it may not approach to the perfection
of Seraphims, and that of Mr. Boyle, let it be as near as it can, and we
shall not account ourselves amongst the most unhappy, for having
made some virtuous addresses to that fair sex."[8]

We need to take these arguments for chastity seriously because the
necessity of chastity for undistracted devotion to God is deeply connected in Boyle's texts to its necessity for proper investigation of nature. Boyle held experimental philosophy to be the highest form of
devotion to God, and inasmuch as celibacy is necessary for undistracted devotion, it is ipso facto necessary for experimental philosophy. Serving God through experimental philosophy is best done by a
celibate man.

His basic argument is that God's creatures provide representations
of the divine attributes available for our observation, representations
which "would very assiduously solicit us to admire him, did we but
rightly discern him in them" (*W*, II, 62). Thus, the man who understands God's creatures is in the best position to prove God's existence,
wisdom, and providence. In fact, such a man becomes a priest. In
*Some Considerations Touching the Usefulness of Experimental Natural Phi-
losophy,* Boyle quotes Philo's claim that "The whole world is to be accounted the chiefest temple of God; the *Sanctum Sanctorum* of it is of
the purest part of the universe, heaven; the ornaments, the stars; the
priests, the ministers of his power, angels and immaterial souls." And,
Boyle argues, qualified men are priests in that temple: "it would not
be rash to infer that if the world be a temple, man sure must be the
priest, ordained (by being qualified) to celebrate the divine service not
only in it, but for it" (*W*, II, 32). But not just anyone is a priest in and
for the natural world; one is ordained to this priesthood by being
qualified, and one is qualified by virtue of one's knowledge of nature.
"[M]en," Boyle claims, "were born with a right to priesthood; so reason is a natural dignity, and knowledge a prerogative, that can confer
a priesthood without unction or imposition of hands" (*W*, II 8). Boyle's

contemporaries recognized the priesthood of scientists, men devoted to the exploration of nature for its revelation of God, and they understood Boyle to be such a priest; "Sir, you are by divine endowments consecrated a chief in that priesthood," Beale wrote to him in 1663 (*W,* VI, 341–342). The laboratory had become the place of worship; the scientist, the priest; the experiment, a religious rite.

We have found that male chastity allows men to serve God undistractedly through experimental science, whereas female chastity serves male chastity; when we find that women's modesty serves the same purpose, we are ready to formulate a basic principle of Boyle's gender economy: Womanly women are chaste and modest, and serve men who, as manly, are chaste and modest experimental philosophers.

Boyle's works reveal that both men and women are to be modest, but modesty in men and modesty in women are two very different things. For women, modesty means modesty of the body and is contrasted with boldness, impudence, and wantonness. Whores are impudent; women are modest. And once again, immodesty in women turns out to be a problem for men because it tempts them. In a letter addressed only to "Madam," Boyle claims that Cleanthe's immodesty (her "masculine boldness") makes her guilty of kindling fires even though she quenches them (BP, XXXVII, 152). This was the letter in which he commented that thousands would be whores if they could get away with it. Boyle's solution was not that women be bashful, but, he said, "we require . . . a modest reservedness of looks and gestures that countenances not vice and such as may quench all un[w]arrantable flames in the kindling, silences all discourses that do but glance at immodesty" (BP, XXXVII, 153). Women's modesty, like their chastity, serves men's chastity by quenching illicit male desire and silencing speech that might express or give rise to it.

Whereas in Boyle's view female modesty is of the body, male modesty is modesty of the mind, a virtue of the scientist; it is one of the paramount requirements for the new experimental philosopher as he sets out to produce facts.[9] Boyle argued that to count as a fact, a phenomenon had to be reliably witnessed. Therefore, experiments had to be public, or at least appear to be public, in order to be counted reliable evidence for facts. Merely performing an experiment did not guarantee a matter of fact; experimental performances were supposed to be "attested by eyewitnesses," and the best way to get witnesses was to perform the experiments in a social space. Air-pump experiments, for example, were often performed before members of the Royal Society at their meetings. Of course, it was impossible for everyone in the experimental community to be present at experiments; therefore, reports of experiments were of supreme importance in producing facts. The man who wrote up an experiment for the reading public had to be trusted, and modesty was one of the most important characteristics of the trustworthy experimenter. Shapin and Schaffer argue that a virtuoso could display his modesty and hence his credibility in a number of ways. The modest man wrote lab reports; that is, he reported experimental trials piecemeal, as opposed to the confident man who wrote systems of natural philosophy. Men who wrote lab reports were, in Boyle's words, "sober and modest men," "diligent and judicious" philosophers, who did not "assert more than they [could] prove" (Shapin and Schaffer, *Leviathan,* 65).

The modest man also wrote in a "naked way" as opposed to a florid way; he adopted the "masculine style" advocated by Thomas Sprat in his *History of the Royal Society.* Briefly, in this matter of style, gender is transformed into national character, travels abroad, and returns to England transformed yet again, this time into genre. "Eloquence," Sprat exclaims, "ought to be banished out of all civil societies." It is a cause of corruption.[10] Eloquence, ornament in speech, and flowery style he attributes to Europeans, especially Italians and French:

> There have, 'tis true, of late, in many parts of Europe, some
> Gentlemen met together, submitted to Common Laws, and
> form'd themselves into Academies. But it has been, for the
> most part, to a far different purpose [from that of the Royal
> Society]: and most of them only aim'd at the smoothing of
> their Style, and the Language of their Country. Of these the
> first arose in Italy; where they have since so much abounded,
> that there was scarce any one great City without one of these
> combinations. But that, which excell'd all the other, and kept it
> self longer untainted from the corruptions of speech, was the
> French Academy at Paris. (Sprat, *History,* 39)

Though they kept themselves longer untainted than the Italians, the French, too, yielded and were corrupted. But the men of the Royal

Society, Sprat assures us, maintain "a constant Resolution, to reject all the amplifications, digressions, and swellings of style: to return back to the primitive purity, and shortness, when men deliver'd so many things, almost in an equal number of words. They have exacted from all their members, a close, naked, natural way of speaking; positive expressions; clear senses; a native easiness: bringing all things as near the Mathematical plainness, as they can" (113). Moreover, Sprat argues,

> if ever our Native Tongue shall get any ground in Europe, it must be by augmenting its Experimental Treasure. Nor is it impossible, but as the Feminine Arts of Pleasure, and Gallantry have spread some of our Neighbouring Languages, to such a vast extent: so the English Tongue may also in time be more enlarg'd, by being the Instrument of conveying to the World, the Masculine Arts of Knowledge. (129)

Abroad one learns flowery speech and the concomitant feminine arts; in England, from the new experimental science, one learns plain speaking and masculine arts of knowledge. This masculine knowledge, produced through experimental methods, should be expressed in a manly style and set down in a lab report.

Finally, the modest man displayed his modesty by keeping to the safe side of the boundary between matters of fact and causal hypotheses or theoretical explanations of those facts. Boyle wrote to his nephew that "in almost every one of the following essays I . . . speak so doubtingly, and use so often, *perhaps, it seems, it is not improbable,* and such other expressions, as argue a diffidence of the truth of the opinions I incline to, and that I should be so shy of laying down principles, and sometimes of so much as venturing at explications." Here theoretical modesty is contrasted with empirical boldness: "I dare speak confidently and positively of very few things," Boyle says, "except matters of fact" (*W,* I, 307; Shapin and Schaffer, *Leviathan,* 67).

We might note that modesty is still considered a virtue for experimental scientists, one that can be essential in helping to establish scientific facts. Investigating the interaction of theoreticians and experimenters in the development of solar neutrino physics, Trevor Pinch focused on the collaboration between research chemist Ray Davis and astrophysicist John Behcall.[11] In 1965, Davis built an experimental apparatus (roughly, an Olympic-sized tank of cleaning fluid a mile down the Homestake gold mine in Lead, South Dakota) to test Behcall's calculations of the solar neutrino flux. The experiments produced what scientists came to refer to as the "solar neutrino anomaly"; that is, the detection rate of solar neutrinos turned out to be lower than theory predicts (Pinch, "Theoreticians," 79). Davis's experiments clashed with Behcall's theoretical predictions, yet the experiments have maintained their credibility. This is in contrast to the norm. Usually when experiments clash with theory, the experiments lose credibility. And, indeed, although Davis's fellow research chemists and other astrophysical neutrino experimenters found his work unobjectionable, the theoretical nuclear astrophysicists "had little to lose" if the experiment were discredited. The theoretical domain would in fact be spared all embarrassment if the experiment could be shown to be faulty. Nevertheless, even though Davis's experiment has received friendly criticism from theoreticians, Pinch notes that most theoreticians now stand by it, not least because of Davis's modesty. "In giving reasons why they believe Davis, many draw attention to his modesty. By this they mean that he has not made any great theoretical claims for his results." Davis thus offers a fine example of Boyle's model scientist; Pinch says of Davis, "[m]any respondents have commented to me that they have been most impressed by Davis [because they see him as acting correctly]. He has all the qualities of an ideal experimenter—he appears to be careful, modest and very open with his results."[12]

In Boyle's view, the theoretically modest man is innocent of even the knowledge of theoretical explanations. Boyle claimed that he "could be very well content to be thought to have scarce looked upon any other book than that of nature," and again, perhaps a bit disingenuously, "I had purposely refrained from acquainting myself thoroughly with the intire system of either the Atomical, or the Cartesian, or any other whether new or received philosophy." He avoided, he said, a systematic acquaintance with the work of Gassendi, Descartes, and even Bacon, that he "might not be prepossessed with any theory or principles" (W, I, 317; Shapin and Schaffer, Leviathan, 68; see also W, I, 355 and 302).

Boyle's theoretical modesty and innocence not only allow the constitution of matters of fact, they also constitute an epistemologically modest and innocent masculinity, opposed to a corporeally modest femininity. Both are necessary for the production of facts: men's modesty directly contributes to the production of facts, while women's modesty indirectly contributes, by insuring that they do not distract men from divine service.

It is instructive to compare Boyle's production of genders with another, similar achievement to be found in the Rev. William Gouge's *Of Domesticall Duties*, a text whose popularity is indicated by its having been reprinted several times throughout the seventeenth century. Gouge fully intends us to see that his manly man and womanly woman are not only made for each other, but made in contradistinction to one another. Following the dedicatory epistle and before the table of contents, he sets out a table displaying womanly characteristics in the left column, each carefully opposed to the corresponding manly quality in the right column. We find, for example, that the duties of a wife are summed up under the "dutie of subjection," while the duties of a husband comprise wisdom and love; she must have "an inward wiue-like feare" and "an outward reuerend cariage towards her husband, which consisteth in a wiue-like sobrietie, mildnesse, curtesie, and modestie in apparell," while he has "an inward intire affection," and "an outward amiable cariage towards his wife, which consisteth in an husband-like grauitie, mildnesse, courteous acceptance of her curtesie, and allowing her to weare fit apparell." She is obedient; he wisely maintains his authority and forbears "to exact all that is in his power."[13] Gouge wants a man who is a pious Christian, a middle-class, heterosexual businessman and ruler. To be a ruler, this man needs a complementary opposite, a wife, in subjection to him. Boyle wants a man very like Gouge's: a pious Christian, a ruling-class, heterosexual man of science.

Both of these manly men, Gouge's businessman and Boyle's scientist, are urban dwellers engaged in civil pursuits and can be understood in contrast to the traditional vision of the manly man engaged in the heroic pursuit of martial arts. Although Linda Woodbridge argues that a civil, urbane masculinity displaced male valor in arms during the early Renaissance, still in the mid-seventeenth century Boyle felt the need to argue that the pursuit of "gowned" valor was as heroic, indeed was more heroic, than heroism on the battlefield.[14]

The anxiety Boyle displayed on the subject is one he shared with other Renaissance and early modern writers. In part because he was celibate, civil, and urbane, the new experimental scientist ran the risk of being effeminate, a "haec vir." Tracing the hic mulier/haec vir controversy from the 1570s through 1620, Woodbridge explains that early writers understood women's transvestism as a deliberate challenge to the immutability of sexual distinctions. And throughout the controversy, we find that when gender characteristics are transferred from one sex to the other, writers worry not only that women become men and men, women, but also that gender proliferates; a third category is created, the "hermaphrodite," "androgine," or "hic mulier" (man-woman), and a fourth, the "haec vir" (woman-man). And although many writers treated transvestism and the concomitant behavioral changes favorably, others referred to them as "monstrous." Thus in 1576 Phillip Stubbes stated that "these Women may not improperly be called *Hermaphrodita,* that is, Monsters of bothe kindes, half women, half men." And in 1619 the Rev. John Williams preached that God "diuided male and female, but the deuill hath ioyn'd them, that *mulier formosa,* is now become *mulier monstra superne,* halfe man halfe woman."[15]

Woodbridge notes that "[w]hether praising or damning, comment on women in masculine attire was almost always accompanied by remarks on male effeminacy" (*Women,* 141). Paradigmatically, *Hic Mulier,* the 1619 tract attacking the man-woman, was followed immediately by *Haec Vir,* a work that turns on effeminacy in men. Difference between the two sexes is certainly one of the issues here, but superficial questions about whether women and men could adopt one another's clothes and social roles express a deeper anxiety over the mutability of gender and the question whether it is, after all, a natural category. The author of *Hic Mulier* worries explicitly because "the new *Hermaphrodites"* are "so much man in all things, that *they are neither men, nor women,* but iust good for nothing." And even though, as Woodbridge notes, the author of *Haec Vir* "officially endorses" a binary gender system, s/he allows Hic Mulier to repudiate that "custome" lying behind dress and behavior codes and, by implication, behind gender distinctions.[16]

Boyle uses two tactics to manage the instability and fluidity of gender revealed by effeminacy and mannishness. The new experimental scientist is constituted a heterosexual man by the woman for whom he has a nearly overwhelming, distracting desire. This woman has breasts like Etna, threatening to erupt hot sexuality everywhere. To be a scientist, he cannot allow her to distract him from his work, but in order for him to be a manly man, she must be desirable and he must desire her.

The work of constituting the scientist as a man by making him distinct from (and desirous of) a woman is particularly difficult when the man is constituted in part by qualities, like chastity and modesty, belonging stereotypically to women. To maintain a clear, binary distinction between men and women, Boyle had to show that when modesty belongs to men, it is no longer womanly, but manly; that is, he had to show how a manly modesty differs from the womanly version of that quality. Appropriating chastity for men was more complicated. He had to distinguish manly from womanly chastity, but because he interpreted chastity to mean celibacy, he also had to defend it from strong attacks on celibacy as a papist "devilish practice." Arguably, for Gouge, a single man is not really a man; certainly he is less of a man than his married brothers. Though Gouge never states that everyone who can should marry, his arguments tend that way. He exempts from marriage only those who are impotent, those who have contagious diseases, and those to whom the gift of continency is given. "[T]he Apostle saith indefinitly of all, without exception of any, *to auoid fornication let every man haue his owne wife, and let euery woman haue her owne husband,*" he says. To the objection that some have lived a single life for the kingdom of heaven's sake, Gouge responds, "That is spoken of some particular persons to whom the gift of continency was giuen: not of any distinct conditions, and callings, as if all and euerie one of this or that calling had so done or were able so to doe." And he launches upon a tirade against "the impure and tyrannicall restraint of the Church of Rome, where by all that enter into any of their holy orders, are kept from mariage." The effects of this "Diabollicall doctrine" have been frightening, "as fornication, adulterie, incest, Sodomie, buggerie, and what not. . . . Deuillish must that doctrine needs be, which hath such deuillish effects" (*Duties,* 184–185). Gouge might be speaking directly to Boyle, who makes precisely this objection in his attempt to persuade Lindamor to turn his passion from Hermione to a more "glorious object."

We have seen that Boyle successfully produces a new man of science, and he does his work in such a way that when they become manly qualities, chastity and modesty take on greater value. Nothing is more important than divine service; the experimental philosophy epitomizes divine service; chastity-as-celibacy allows undistracted service through science, while a manly modesty is essential to one's construction of facts through experimentation. Manly chastity and modesty are qualities of the person engaged in the highest calling; womanly chastity and modesty are qualities of the person who keeps out of his way.

It certainly seems that Boyle has left no place for woman in the new experimental science. She hasn't the qualities of the experimental scientist; her very presence threatens to be distracting; she is better excluded. In one sense, this is all quite true. Women were not members of the Royal Society, and were, among others, barred from its meetings. In Sprat's "An Abstract of the Statutes of the Royal Society," we find that "The ordinary meetings of the Royal Society shall be held once a week, where none shall be present, besides the Fellows, without the leave of the Society, under the degree of a Baron in one of His majesties three Kingdoms, or of His Majesties Privie Council; or unless he be an eminent Forreigner, and these only without the leave of the President."[17] Women were not among those qualified to make knowledge by witnessing experiments and attesting their validity.[18] Nor, judging from the minutes of the Royal Society meetings, do women appear to have been a topic for discussion among the men of science.

In another sense, however, Boyle's womanly woman has everything to do with the new experimental science and is everywhere present in her absence. This is well expressed by saying that she is "at the margins" of science: she has to be there to form a contrast, to give shape to the portrait of the experimental man, but it is easy to overlook or miss her, since, like many of the objects around the edges of the picture, she frames it and doesn't appear to be in it. Boyle rarely mentions the presence of women at experiments, but in one of the passages where he does mention them, he offers us a vignette that nicely captures his construction of the womanly woman at the margins of experimental science, serving as a foil for the new experimental scientist. In *New Experiments Physico-Mechanical Touching the Spring of the Air,* a work describing experiments with the air-pump, Boyle tells us that, having divers times tried the experiment of killing birds in a small receiver, we commonly found, that within half a minute of an hour, or thereabout, the bird would be surprised by mortal convulsions, and within about a minute more would be stark dead, beyond the recovery of the air, though never so hastily let in. Which sort of experiments seem so strange, that we were obliged to make it several times, which gained it the advantage of having persons of differing qualities, professions and sexes (as not only ladies and lords, but doctors and mathematicians) to witness it. And to satisfy your Lordship, that it was not the narrowness of the vessel, but the sudden exsuction of the air that dispatched these creatures so soon; we will add, that we once inclosed one of these birds in one of these small receivers, where, for a while, he was so little sensible of his imprisonment, that he eat very cheerfully certain seeds that were conveyed in with him, and not only lived ten minutes, but had probably lived much longer, had not a great person, that was spectator of some of these experiments, rescued him from the prosecution of the trial. Another bird being within about half a minute cast into violent convulsions, and reduced into a sprawling condition, upon the exsuction of the air, by the pity of some fair ladies, related to your Lordship, who made me hastily let in some air at the stop-cock, the gasping animal was presently recovered, and in a condition to enjoy the benefit of the ladies compassion. And another time also, being resolved not to be interrupted in our experiment, we did at night shut up a bird in one of our small receivers, and observed that for a good while he so little felt the alteration of the air, that he fell asleep with his head under his wing; and though he afterwards awaked sick, yet he continued upon his legs between forty minutes and three quarters of an hour: after which, seeming ready to expire, we took him out, and soon found him able to make use of the liberty we gave him for a compensation of his sufferings. (*W,* I, 106–107)

This rich passage describing a series of experiments designed to discover the relationship between the very low pressure created in the air-pump and respiration reveals the interdependence of Boyle's gender achievements and his scientific achievements. In writing up the experiments, Boyle presents himself as modest by adopting the plain, manly style and writing a lab report instead of an ornate theoretical treatise. As we have seen, this masculine style lends credibility to Boyle as he calls his readers to witness his experiments vicariously and to attest their validity and confirm the facts they warrant. Moreover, as he establishes scientific fact in the passage, he displays the basic rules for good experimental procedure: for example, experiments should be careful and precise, and this passage reveals the necessary precision by constantly mentioning the time it takes the experimental animals to expire. To insure that the results are unimpeachable, the experiment should be run more than once and should be done in the presence of witnesses. When the results are strange or unexpected, a variety of witnesses is advantageous for securing credibility.

Who are these witnesses? Boyle's list indicates that he takes class and gender to be among the important categories: the witnesses mentioned are "persons of differing qualities," including both aristocrats ("ladies and lords") and bourgeoisie (professionals like "doctors and mathematicians"). The aristocratic ladies and lords also exemplify "persons of differing . . . sexes." And witnesses are not limited to any particular profession; they can be doctors, mathematicians, or, Boyle implies, members of any other profession. Prima facie, then, the new experimental science appears to be quite democratic. But let us look more closely; can just anyone witness experiments and make knowledge? In the first place, Boyle does not include in his list the working-class men who were usually present to do the hard labor involved in his experiments. Working the air-pump, turning the handle to pump out the air, became increasingly difficult as the air pressure decreased in the receiver, and in other passages Boyle reveals that this work was often done by, for example, a blacksmith.[19] We may conclude, then, that the persons of differing qualities suitable for witnessing do not include working-class persons.

Boyle explicitly tells us, however, that the ladies "witness" the experiment of suffocating animals; but in what sense do they witness it? When the institution of witnessing was formalized through the creation of the Royal Society, women, who could not be members, were implicitly excluded from the activity. Moreover, ladies' names are never among those Boyle mentions as attesting his veracity.[20] Women and other non-scientists, then, *watch* his experiments; virtuosi, including doctors and mathematicians, *witness* them and attest to the facts produced thereby. Furthermore, Boyle makes it obvious why women merely watch. The women who enter his laboratory are unfit for serious witnessing or experimenting; in fact, their presence turns out to disrupt the experiment altogether.

The danger of sexual distraction and disruption that women present to science quickly appears when we learn who these ladies are and remember their gender qualities. Boyle tells us that the ladies are "related to your Lordship," that is, to Boyle's eldest brother, to whom *New Experiments* was addressed. He is probably referring to their sister, Lady Katherine Ranelagh, and her daughters, but in any case, Boyle always publicly presents his kinswomen as women of "strict virtue," which invariably includes piety, chastity, and modesty. His *Occasional Reflections upon Several Subjects* is prefaced by a letter to "Sophronia"—his sister, Lady Ranelagh—and there he says of her that

> she is wont to persuade piety as handsomely in her discourses,
> as she expresses it exemplarily in her actions; and might, if her
> modesty did less confine her pen to excellent letters, both make
> the wits of our sex envy a writer of hers, and keep our age
> from envying antiquity for those celebrated ladies, who, by
> their triumphant eloquence, ennobled the people of Rome, and
> taught their children to sway those rulers of the world.

"But," he says to her, "I dare not prosecute so fruitful a subject [that is, her attainments], for fear of offending your modesty; since that predominant virtue gives you so great an undervaluation for all your other qualities, that it is as much your custom to look even upon small praises as flatteries, as it is your prerogative to keep great ones from being so" (*W,* II, 324).

Just so, in the letter to his sister Mary, Countess of Warwick, prefatory to *Some Motives and Incentives to the Love of God,* Boyle remarks that

> though I can praise you without either disbelieving myself, or
> fearing to be disbelieved by any that knows you; yet . . . I know
> your piety and your modesty would peculiarly disallow it upon
> this; where the subject I am to entertain you with is of such a
> nature, as would make a flaunting address but a very unsuit-
> able introduction to it. The nature of my theme, as well as the

strictness of your virtue, and our friendship, forbidding me
here to celebrate you, otherwise than by letting the world see,
that I dare, even in a dedicatory address, without fear of dis-
pleasing you, forbear to celebrate you. (*W,* I, 245)

Thus, the ladies who enter Boyle's laboratory have already been presented to us as virtuous women, chaste, modest, and pious. They are
so modest that they do not even want to be praised; praise might em-
bolden them, make them the object of that public attention which for
women always threatens to be the sexual attention of men. As chaste
and modest women, his kinswomen would quench male desire in the
kindling, and silence "all discourses that do but glance at immodesty."
Surely, then, these women can enter the laboratory without distracting the scientist and disrupting his work. But no. Boyle's vignette
shows that they did disrupt the experiment, and so severely that they
forced him to work at night to avoid them. When he tells us that "a
great person" and the "ladies" made him stop the experiments to save
the suffocating birds, Boyle accomplishes two things: he creates and
manages a boundary between those who are engaged in scientific
work and those who are not, and he shows that science is not gen-
der-neutral; instead, gender is central to good science.

In this passage, Boyle manages through the same maneuver both
the distinction between scientists and non-scientists and the essential
boundary between facts and causal hypotheses or theories accounting
for the facts. Ostensibly, the important fact produced in this series of
experiments is that "the exsuction of air suffocates animals," and
Boyle interprets the experiments to support this fact as opposed to the
hypothesis that it is "the narrowness of the vessel" that kills them. In
this passage as in many others, however, the unstated claim that "air
is being sucked from the receiver" is tacitly moved across the boundary from theoretical hypothesis to fact. On Boyle's interpretation, had
not the "great person" interrupted the experiment in which a bird
was to be left in a receiver full of air, we would have seen that the
bird would live quite a long time; until he used up the air, in fact.
And had not the "ladies" interrupted an experiment in which the air
was being pumped out of the receiver, the bird, already "cast into violent convulsions, and reduced into a sprawling condition," would
shortly have suffocated. These two "failed" experiments help warrant

both the ostensible and the tacit facts at issue. But neither the "great person" nor the "ladies" intended to discover facts; only the experimenter realizes the important implications of two otherwise unfortunate events and makes a success of failure.

The near-failures crystallize the importance of the boundary between the scientist and the non-scientist. The serious scientist is trying to discover the relation between the very low pressure created in the air-pump receiver and respiration. For him, pursuit of the facts is valued above compassion for animals used to discover the facts. And his devotion to pursuit of the facts is made clear through his attempt to avoid distraction and the disruption of his work by experimenting at night; he is "resolved not to be interrupted." Non-scientists, particularly compassionate "fair ladies," however chaste and modest, upset the necessary hierarchy of values and disrupt the pursuit of knowledge. The quality of compassion is good and Boyle exhibits it himself in this passage, but in the man of science it takes its proper place after the serious pursuit of knowledge; only after the experiment is successfully completed does a scientist liberate his suffering experimental animals.

In this vignette, Boyle throws the experimental scientist into sharp relief by contrasting him with womanly women. The best of women, pious, chaste, modest, and compassionate, are rendered unfit for science by the very qualities that make them the best of women. But these same qualities, when they belong to a man, make him the best of men, pious, chaste, modest, and compassionate, and admirably fit him for experimental science. A woman must make every attempt to control her desirable sexuality through chastity and modesty. But if she were not "there" and desirable, the manly experimenter would not need to be chaste, to overcome his desire for the sake of divine service through experimental science, and would not pursue the facts with manly modesty and proper compassion.

2
Now We Don't

The production of Boyle's Law in 1661 was closely tied up with several questions, as vexing as they were pressing, for natural philosophers in the early seventeenth century. To twentieth-century students of science, the most familiar of these is the debate over the possibility of a vacuum; most Aristotelians held that nature abhors a vacuum, while others, particularly proponents of what was to become a rival paradigm, viz., mechanism, held that one was possible. Opposition on this question did not, however, follow strict party lines; for example, Descartes, that arch-mechanist, was a plenist, i.e., he rejected the possibility of a vacuum. And not all hylozoists, who believed that matter can move itself or that Nature is a Spirit who infuses and guides the world, thought that a vacuum was impossible. For example, Campanella, that arch-animist, argued that although "the common sense of things abhors a vacuum," it is possible to create one by violence.[1]

Generally speaking, seventeenth-century mechanists rejected the view that matter is animated and sought explanations for water pumps, siphons, and so on, according to which water and other matter is "brute" or "stupid," as they put it, and blindly follows certain laws set out by God at creation. There were many issues at stake in this dispute: which was the best account of phenomena like water pumps? Since these accounts were part of different world views or paradigms, which paradigm was best? And to the extent that the Catholic Church had put its authority behind an Aristotelian world view and Aristotelian accounts, many felt that the authority of the church was at stake in the dispute.

The question concerning the possibility of a vacuum was raised sharply by experiments designed to test directly whether a vacuum could be produced, and by those designed to test whether the air has weight or density and whether it has "elater" or "spring." Later, these questions were compressed into the question whether the air has pressure. Galileo argued that air weighs nothing and does not exert pressure on objects; but his compatriot, Giovanni Batista Baliani, believed that the air does have weight and that "the higher we go in the air the less heavy it is." Isaac Beeckman, too, argued that the air has weight; the upper air weighs down the lower air, compressing it like a sponge. He argued that a vacuum is possible, and when a space is emptied, for example in a water pump, water rushes into it both because of its own weight and "on account of the immense depth of the superincumbent air" pressing down on it and forcing it into the space. Galileo's alternative explanation was that the vacuum exerts a force of attraction upon the water, though this force is limited; thus, at about the height of nine or ten meters from the ground, water pumps no longer work because the vacuum is unable to exert any more force and the column of water breaks under its own weight.[2]

Do suction pumps work because nature abhors a vacuum or because air has weight and spring? Already we can see that the problem was to find a crucial experiment providing data which one hypothesis, but not the other, could account for. Torricelli explains one such experiment performed, he says, "not simply to produce a vacuum, but to make an instrument which might show the changes of the air, now heavier and coarser, now lighter and more subtle." If he could find the (real) cause of the resistance that is felt in trying to produce a vacuum, it would be useless to say that the vacuum itself resisted its own production. "We live," Torricelli argues, "submerged at the bottom of an ocean of elementary air, which is known by incontestable experiments to have weight" (quoted in Middleton, "Torricelli," 19–20). Torricelli takes the following experiment, then, to be a crucial one and interprets it to mean that the weight of the air causes the resistance to a vacuum, so Galileo is wrong to think that the resistance we feel results from the force of the vacuum itself, i.e., from a force of attraction, and it is wrong to think that Nature abhors a vacuum.

Torricelli (or more probably his friend Vincenzo Viviani) took a glass tube about 43 inches long, open at one end and closed at the other, and filled it to the brim with mercury; covering the open end with a finger, he turned it upside down and thrust it into a bowl half full of mercury. (See Figure 1.) When the finger was removed, the column of mercury in the tube dropped so that it stood about 27 inches above the surface of the mercury in the bowl. To prove that "nothing took the place of the quicksilver [mercury] in the [glass tube] that was being emptied," i.e., that there was a vacuum in the space at the top of the tube, he says,

> the basin was filled up to [the brim] with water [i.e., water was poured on top of the mercury in the basin]; and on raising the [tube] little by little, it was observed that when the mouth of the [tube] reached the water, the quicksilver in the neck came down, and the water rushed in with horrible violence and filled the [tube] completely up [to the top].

Torricelli states his explanatory hypothesis clearly:

> On the surface of the liquid in the basin presses a height of fifty miles of air; yet what a marvel it is if the quicksilver enters the glass [tube], to being in which it has neither inclination nor repugnance, and rises there to the point at which it is in equilibrium with the external air that is pushing it!

We should also note his comment that he cannot after all use his apparatus to find out when the air is coarser and heavier and when more subtle and light because "the level changes from another cause (which I never thought of), that is, it is very sensitive to heat and cold, exactly as if the [space at the top of the tube] were full of air" (quoted in Middleton, "Torricelli," 19–21).

The conclusion that Torricelli drew from this experiment, to be repeated throughout Europe for many years after, was hotly disputed. In fact, the possibility of a vacuum had been disputed since Aristotle's response to the Democritean atomists. The atomists had distinguished two sorts of vacuum. In one sense they argued that any given space could be completely empty but for the atoms which exist in space. Bodies (including atoms) are only able to move because of the empty spaces around them into which they move; if there were a plenum, motion would be impossible. In the second sense, there is also empty space, the interstitial vacuum, between the atoms that make up large bodies; and when a body expands, the dispersed vacuum or empty spaces between its atoms increase, while in contraction the atoms come closer together and the spaces between them decrease.

Aristotle gave a number of arguments against the atomists designed

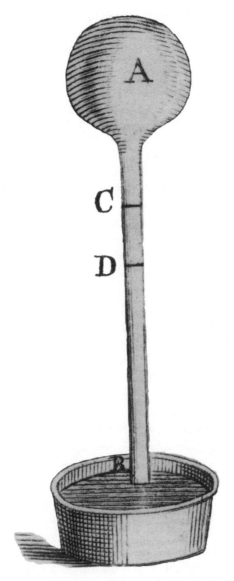

Fig. 1. The Torricelli Tube inverted into a basin of mercury. This tube has a bulb at the end suitable for including experiments such as the carp's bladder. From *W,* I, Plate 2.

Reproduced with permission of the Department of Special Collections, Stanford University Libraries.

to show that there is no empty space occupied by a body, nor is there empty space within a body; here we will examine only three. Motion is possible without empty space, "for bodies may simultaneously make room for one another, though there is no interval separate and apart from the bodies that are in movement." All movement, he suggested, is like the rotation of liquids stirred in a basin, the parts succeeding one another continuously. Second, given the principles of motion according to which a projectile moves because the air it pushes moves around and in turn pushes the projectile "with a movement quicker than the natural locomotion of the projectile wherewith it moves to its proper place," the motion of projectiles in a void would be impossible.

Against the atomists' argument from condensation and rarefaction, Aristotle offered an alternative account of the phenomena of the expansion and contraction of bodies; the density of a body, he said, is one of its qualities just as "hot and cold and the other natural contrarieties" are. Thus, when water turns into air, its quantity of matter does not change, though its volume does. Certain bodies may expand or rarefy and so change their volume just as the temperature of bodies sometimes changes from hot to cold.[3]

Seventeenth-century students of nature still found many of Aristotle's arguments persuasive. René Descartes, arguing for a strictly mechanical physical theory but against the possibility of a vacuum, said that the space at the top of the Torricelli tube was empty of the relatively large particles of air, but filled with "subtle matter," later known as the aether. This subtle matter was fine enough to move through the walls of the glass tube and into the space as the grosser particles of air left it. On the other hand, although he followed Descartes in adopting a version of atomism, Pierre Gassendi disagreed with him over the vacuum and argued that the space might be completely empty.

The problem, of course, was that the space might still be filled with rarefied air. Roberval was of this opinion and he devised a number of experiments to test the expansion of the air, the most dramatic and popular of which was "the experiment with the carp's bladder." Taking the air bladder from a carp, Roberval pressed out the air and tied the bladder closed; it was then put into the closed end of a Torricelli tube, and the tube was filled with mercury and inverted into a basin of mercury. The bladder, now in the space in the top, was observed to have inflated. (See Figure 1.) Roberval interpreted these results to indicate that air is ordinarily compressed by the weight of the "super-incumbent" atmosphere, but when it is released from that weight, as in the Torricelli experiment, it acts like a spring and rarefies sponta-neously.

In 1648, Mercenne expressed the theoretical impasse in his remark to Hevelius that though they multiplied experiments, "nothing comes of it." No crucial experiment could be found to decide whether a vac-uum is impossible or whether the air has pressure (i.e., weight and spring).[4] However, to test the hypothesis that the air has weight and exerts a pressure which decreases as one approaches the top of the atmosphere, Blaise Pascal's brother-in-law, Perier, performed the Tor-ricelli experiment at the base of the Puy-de-Dôme (a mountain in central France), at the top, and at several points on the way to the top. He found that the level of mercury in the tube dropped as he ascended the mountain.

The debate between Gassendi and Descartes was taken up in Eng-land as well. Descartes's plenist views were taken up at Cambridge University, defended and taught by Henry More and Ralph Cudworth, while Gassendi's views were first defended in print by Walter Char-leton in his *Physiologia Epicuro-Gessendo-Charltoniana* in 1654. He cited the Torricelli experiment as support for the existence of the vacuum, but he also explained the rarefaction and condensation of air and other bodies in Epicurean terms: atoms can be compressed by the weight of others above them and come closer to one another, "leaving less intervals, or empty spaces betwixt them than before." And when "the external force that compressed them ceases," the atoms spring apart (and sometimes rebound off one another). This is what happens to the air at the top of the Torricelli tube.[5] Here we find two important additions to the understanding of the air, first, it has spring-like or elastic qualities, and these qualities give rise to, second, its kinetic qualities—its atoms rebound off one another.

Robert Boyle began a series of experiments on the air in the 1650s and joined the dispute with his first publication on the subject in 1660, *New Experiments Physico-Mechanical Touching the Spring of the Air, and its Effects; Made, for the Most Part, in a New Pneumatical Engine.* It was Boyle's "pneumatical engine," one of the first vacuum pumps, that stirred the imagination and catapulted him to the attention of virtuosi throughout Europe, helping to make him, even today, a paradigm of the experimental scientist. Boyle was probably inspired to have an "air-pump" and to make experiments with it by Schottus's 1657 description of Otto von Guericke's "way of emptying glass vessels, by sucking out the air at the mouth of the vessel, plunged under water." Boyle was delighted with von Guericke's experiment because it rendered the "great force of the external air" more obvious and conspicuous than any experiment he had yet seen, and he determined to have a similar apparatus. He set his assistant, Robert Hooke, and London's best instrument maker, Ralph Greatorex, to contrive a pump that was easier to manage than von Guericke's. Hooke finally came up with a successful design.

Boyle's first air-pump is pictured in Figure 2. The pump's piston, or "sucker," as he termed it, worked on a rack and pinion moved by turning the handle of a crank. The cylinder containing the pump rested in a wooden frame and itself supported a glass globe or "receiver" which Boyle had had blown especially for the purpose. The globe had a small opening in the bottom, through which the air passed, and a larger opening in the top, sufficient to admit the experimental apparatus. Between the globe and the pump cylinder was a brass stopcock or key which could be turned to open or close off the globe from the pump. Boyle complained that the workmen were unable to blow a globe as large or of as convenient a shape as he wished, but they finally settled on one that was "tolerably fit, and less unfit than any of the rest" (*W,* I, 6–7).

Boyle carried out his experiments and made his contributions to the debate in terms of the pressure, the "weight and spring," of the air, not in terms of the possibility of a vacuum. He did not want to commit himself to either side of the question whether it is possible to produce an absolute vacuum. Therefore, when he uses the term, he is careful to tell us that by "vacuum"

> I here declare once for all, that I understand not a space,
> wherein there is no body at all, but such as is either altogether,
> or almost totally devoid of air. (*W,* I, 10)

His fellow virtuosi made this distinction between an absolute vacuum and the evacuated receiver by referring to the space inside the receiver as "the Boylean vacuum."

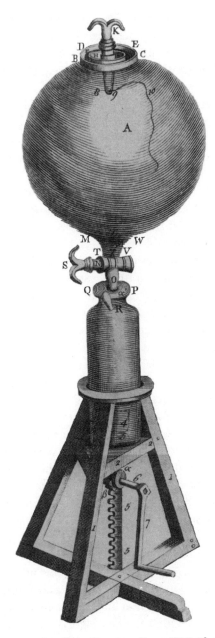

Fig. 2. Boyle's first air-pump. From *W,* I, Plate 1.

Reproduced with permission of the Department of Special Collections,
Stanford University Libraries.

In any case, the air-pump leaked. This presented less of a problem for quick experiments, Boyle tells us, "because the air may be faster drawn out by nimbly plying the pump, than it can get in at undiscerned leaks"; but for those experiments requiring that the receiver be empty of air for long periods of time, the leaks were a serious difficulty. Even a very small leak would ruin the vacuum. However, it is instructive to look at what Boyle says about this problem; he admits that "in spite of all our care and diligence we never were able totally to exhaust the receiver," but argues that "the internal air can be much faster drawn out than the external can get in, till the receiver come to be almost quite empty. And that is enough to enable men to discover hitherto unobserved phaenomena of nature" (*W*, I, 10). This remark indicates, not that Boyle was a careless experimenter, or that he was attempting to pass off low-pressure phenomena as vacuum phenomena, but that he was primarily interested in phenomena at low pressure, or more generally, in the pressure of the air, not in the issue of the vacuum. This interpretation has the virtue of taking seriously Boyle's repeated claims that he was not trying to determine the possibility of an absolute vacuum and his claim that his book was about the spring of the air and its effects. Thus, he explicitly sets out a model of the air according to which it has pressure and he offers the model as the "notion" or hypothesis that explains (and is supported by) most of his experimental results.

According to the explanatory model he adopts, the air near the earth is a "heap of little bodies, lying one upon another, as may be resembled to a fleece of wool." For fleece consists of hairs each of which can, "like a little spring, be easily bent or rolled up; but will also, like a spring, be still endeavouring to stretch itself out again" (*W*, I, 11). When released from compression, the particles of air, like wool fleece, spontaneously dilate; this power of "self-dilatation" is a key notion in explanations of the kinetic effects of the air particles observed in the experiments (and it caused problems for Boyle, as we will see in chapter 14 below). The only element of this model Boyle does not argue for is atomism, i.e., the picture of air as composed of small particles. Each of the others—that the atmosphere extends to a great height, that air has weight, that the weight of the particles above compresses the particles below, and that the lower particles, once released from the great weight of the particles above, dilate and "take up more room than before"—Boyle struggles to establish. Thus, Boyle and other virtuosi spent a great deal of time and energy trying to weigh the air. Boyle describes one such attempt as follows:

> I found a dry lamb's bladder containing near about two thirds of a pint, and compressed by a packthread tied about it, to lose a grain and the eighth part of a grain of its former weight, by the recess of the air upon my having prickt it: and this with a pair of scales, which, when the full bladder and the correspondent weight were in it, would manifestly turn either way with the 32^d part of the grain. (*W*, I, 13)

The precision of his scales allowed Boyle reasonably to conclude that two thirds of a pint of air weighed about one and one eighth grain.

It is not difficult to appreciate why these attempts to weigh the air were the butt of much humor at the time. Thomas Shadwell's *The Virtuoso* satirized Boyle as Sir Nicholas Gimcrack:

> **Longvil:** But to what end do you weigh this Air, Sir?
> **Sir Nicholas Gimcrack:** To what end shou'd I? To know what
> it weighs.
> O knowledge is a fine thing.

Samuel Butler satirized the group of virtuosi who later formed the Royal society:

> Their learned speculations,
> And all their constant occupations
> To measure and to weigh the air
> And turn a circle to a square.

And Charles II himself was said by Pepys to have "mightily laughed" at his protégés in the Royal Society "for spending time only in weighing of air, and doing nothing else since they sat."[6]

Calling upon the estimates of the height of the atmosphere by Kepler (eight miles) and others such as Ricci (fifty miles), Boyle argues that "a column of air, of many miles in height" leans upon corpuscles of air here below. And given that the air has weight, he further argues that the hypothesis of the air's spring explains phenomena such as Pascal's Puy-de-Dôme experiment, in which the Torricellian experiment was "tried at the foot, about the middle, and at the top of that high mountain." In fact, we find that this is Boyle's argument strategy, not only for the hypothesis that the air has spring, but for most of the sub-hypotheses constituting the "corpuscular philosophy," as he referred to the new atomic theory: we should accept the new theory because it is sufficient to explain the phenomena and it has fewer conceptual problems than its rivals; it is, he often says, "more intelligible."

In the first edition of *New Experiments* Boyle discusses forty-three experiments performed with his pneumatical engine; the new standard of care and precision exhibited in those experiments as well as the style and detail of his discussion of them have made Boyle a model experimenter for students. We are not surprised then that James Conant chose some of Boyle's early pneumatic experiments for the first of the Harvard Case Histories in Experimental Science. The case histories are intended to acquaint the novice with historical examples of good scientific methodology so that he or she will understand modern science. Pride of place goes to Experiment XVII from *New Experiments*. Conant explains that one of the earliest objections raised against Torricelli's experimental results was that if the air exerted pressure on the dish of mercury sufficient to force the mercury in the column up to 30 inches, then covering the dish and thereby sealing it off from the pressure exerted on it should cause the mercury in the column to fall; but when the surface of the dish is sealed off, the mercury column does *not* in fact fall. Therefore, Torricelli's broad working hypothesis—that "the earth is surrounded by a sea of air that exerts pressure"—was wrong. As did Torricelli, Boyle answered this objection by arguing that when the dish was enclosed, the pressure on the enclosed mercury remained the same as it was before. But Boyle himself sought to test Torricelli's hypothesis by finding a way to remove the air from the enclosed space around the dish of mercury. To this end, he made use of his air-pump or pneumatical engine.

On Conant's account, Boyle set out to test Torricelli's broad working hypothesis concerning the sea of air by deducing from it a limited working hypothesis susceptible of experimental test:

> When Boyle built his engine he set out to test one deduction from Torricelli's broad hypothesis. When he had his apparatus all arranged he reasoned somewhat as follows: "If I now operate the pump, then the mercury column in the Torricellian

tube should fall." . . . He proceeded with the experiment and the results were as predicted. He employed a similar "if . . . then" type of statement when he considered introducing air into the receiver; the mercury rose, as predicted.[7]

The experiment can be summarized as follows. A glass tube about a yard long and sealed at one end was filled with mercury and inverted into a small cylindrical container half filled with mercury. After the mercury had settled, the height of the column of mercury was marked (by pasting a piece of paper to the tube) and the cylinder and tube were let down into the receiver. The cover of the receiver, with a hole in the middle to pass over the tube, was put in place and all the leaks stopped with melted wax. Boyle remarks that upon closing the receiver "there appeared not any change in the height of the mercurial cylinder, no more than if the interposed glass-receiver did not hinder the immediate pressure of the ambient atmosphere upon the inclosed air; which hereby appears to bear upon the mercury, rather by virtue of its spring than of its weight." The piston was drawn down (by turning the crank) and, Boyle says, "immediately upon the egress of a cylinder of air out of the receiver, the quicksilver in the tube did, according to expectation, subside: and notice being carefully taken (by a mark fastened to the outside) of the place where it stopt, we caused him that managed the pump to pump again, and marked how low the quicksilver fell at the second exsuction." Even after pumping for a quarter of an hour, the mercury did not descend to the level of the mercury in the container, because, Boyle says, enough air leaked into the receiver to keep the mercury up somewhat.

The stopcock was opened and some air let in, and the mercury began to ascend until the stopcock was closed, whereupon the mercury "immediately rested at the height which it had then attained." Boyle goes on to describe further steps in the experiment. But this step served, Boyle says, to satisfy them that the mercury falls until it reaches a state of equilibrium with the external air, the equilibrium being a function of the weight of the mercury and the pressure of the air (W, I, 33–34).

Boyle remarks that he had hoped to "give a nearer guess" at what he refers to as "the proportion of force" between the weight of the mercury and the pressure of the air. He was able to determine the

volume of air in the receiver at atmospheric pressure, the volume of
air that could be contained in the cylinder of the pump and removed
from the receiver at each exsuction, and the weight of precisely the
amount of mercury it took to balance the amount of air removed from
the receiver at each exsuction. The problem was that the volume of
air in the receiver varied both with the level of mercury in the dish
at the bottom and with the diminishing amounts of air removed by
the pump as less air filled the pump's cylinder with each exsuction.
We can sympathize with Boyle's disappointment:

> Because of these . . . and some other difficulties, that require
> more skill in mathematics than I pretend to and much more
> leisure than my present occasions would allow me, I was will-
> ing to refer the nicer consideration of this matter to some of
> our learned and accurate mathematicians, thinking it enough
> for me to have given the hint already suggested. (*W,* I, 36)

In any event, the relationship between the volume of air and its pressure was not determined by experiments with the air-pump. As ex-
citing and innovative as the pneumatical engine was, Boyle did not
use it for deriving the law that bears his name.

It was the publication of *New Experiments* in 1660, reporting the
series of experiments with the air-pump, that gave rise to the controversy which led to the production of Boyle's Law. There were many
responses to Boyle's book, but the three Boyle in turn responded to
were Thomas Hobbes's *Dialogus physicus* and Franciscus Linus's *Tractatus de corporum inseparabilitate,* both published in 1661, and Henry
More's *Antidote against Atheism,* which appeared in its third edition in
1662. All three attacked Boyle's interpretations of the air-pump experiments.

Hobbes was not an Aristotelian, though he was a plenist; he flatly
denied that even a "Boylean vacuum" had been produced by the air-
pump and offered reasonable alternative interpretations of the air-
pump phenomena. Here we will discuss only two. Hobbes conceived
all matter to be infinitely divisible, so that the parts of air, for example, are infinitely subtle. This allowed him to explain why no vacuum
of any sort is created by the air-pump: when the sucker is drawn
down, he says, it displaces air outside the receiver; each part of the
air pushes on the parts adjacent "so that of necessity the air is forced

into the place left by the sucker." It leaks back into the receiver through the spaces between the piston and the walls of the brass cylinder around it.[8]

The Torricelli experiment he explains in the same way: when the tube full of mercury is inverted into a basin of mercury, the mercury descends into the basin, thereby pressing against the air lying above the surface of the mercury in the basin; because that air has nowhere else to go, it either penetrates back through the mercury or leaks back between the mercury and the inside walls of the tube, thus again filling the space at the top of the tube. In any case, Hobbes denies that the air has spring on the grounds that nothing moves itself; he argues that unless we assume that it does, which is impossible, Boyle's model of the air as consisting of little spring-like particles is inadequate. He correctly points out that "you [Boyle] suppose that the air particle, which certainly stays still when pressed, is moved to its own restitution, assigning no cause for such a motion, except that particle itself." Here Hobbes presents himself as a better mechanist than Boyle; he refuses the mysterious notion of the "spontaneous dilation" of the air when released from compression and demands a completely mechanical explanation. To him, the notion of spontaneous dilation is no better than the Aristotelian notion that nature abhors a vacuum (Shapin and Schaffer, *Leviathan*, 363–364).

Franciscus Linus, however, objected to Boyle's work on Aristotelian grounds, arguing specifically in behalf of nature's horror of the vacuum; unlike Hobbes, he offered non-mechanical explanations of the relevant phenomena. It was in his response to Linus that Boyle developed the first version of the Ideal Gas Law.

In the first place, Linus went out to a convenient mountain and performed Pascal's Puy-de-Dôme experiment. He reported that he found no difference in the height of the mercury in the Torricelli tube at the foot and at the top of the mountain, which supported his view that it is not the pressure of the air that keeps up the level of mercury in the tube. Results such as Pascal's, he suggested, were probably due to changes in temperature between the foot of the mountain and the top. He did not suggest, as does Conant, that Perier cooked the data when he performed the experiment for Pascal.[9]

Experiments with the Torricelli tube, including those performed in the air-pump and those performed outside the air-pump, could best be explained, according to Linus, by the action of a funiculus attached to the surface of the mercury and to the top of the tube. In mercury experiments, this funiculus is probably made out of mercury vapor or rarefied mercury left behind as the mercury descends in the tube; as the air in the top of the Torricelli tube becomes more and more rarefied (and Linus held an Aristotelian account of rarefaction), the funiculus, acting rather like a membrane, thickens and contracts all the more, holding up the level of mercury. The membrane does not have unlimited power, however, so that in the Torricelli experiment the mercury stands at 29½ inches because the membrane is unable to pull it up any higher.

Linus offered as direct experimental evidence of the funiculus the fact that when one fills a tube (open at both ends) with mercury, covers each opening with a finger, immerses one end of the tube in a basin of mercury, and releases the immersed finger so that the mercury descends into the basin, one feels the finger covering the top end of the tube being pulled or sucked into the tube. According to Linus, this finger is pulled by the funiculus, and so strongly that he was able to lift up the tube, mercury and all, simply by lifting his finger. This, he said, contradicts Boyle's hypothesis that the external air presses down on the exposed surface of the mercury and pushes the column of mercury up in the tube.

It also explains why it is more and more difficult to ply the pump as the air in the receiver is increasingly rarefied: the increasingly rarefied air itself forms a membrane that attaches to the top of the receiver and pulls against the sucker as it descends in the brass cylinder. And it easily explains Experiment XVII, in which the Torricelli experiment is performed inside an empty glass globe. Here the Torricelli apparatus is inside the receiver of the pneumatic engine, and Linus argues that when the pump's piston is drawn down, the membrane formed by the increasingly rarefied air attaches to the surface of the mercury in the basin and, as it "vehemently contracts," it pulls the surface of the mercury up and the mercury in the tube falls down.[10]

We should note, finally, that Linus did not deny that the air has weight or that it has spring; he just denied that it has enough spring to force the mercury up in a Torricelli tube or to account for any of the other phenomena Boyle used it to explain. As we shall see, this gave Boyle the opening he needed to respond to Linus.

To show that Linus's funicular hypothesis is unnecessary, Boyle argues that as the mercury descends, the air in the top of the tube is rarefied and exerts little pressure, so that one's finger is pressed into the tube by the air above it, not pulled into the tube by a funiculus (*W*, I 129).

The experiment itself, of course, does not allow one to decide between Boyle's hypothesis and Linus's; therefore, Boyle casts about for a crucial experiment that will determine whether the experimental phenomena are caused by the pressure of the air or the pull of the funiculus. The Puy-de-Dôme experiment is, he says, just the one. It appears, Boyle says,

> that the quicksilver being carried up towards the top of the
> atmosphere, falls down the lower, the higher the place is
> wherein the observation is made: of which the reason is plain
> in our hypothesis, namely, that the nearer we come to the top
> of the atmosphere, the shorter and lighter is the cylinder of air
> incumbent upon the restagnant mercury; and consequently the
> less weight of cylindrical mercury will that air be able to coun-
> terpoise and keep suspended. And since this notable phaeno-
> menon does thus clearly follow upon ours, and not upon our
> adversaries hypothesis; this experiment seems to determine the
> controversy betwixt them. (*W*, I, 151)

The funicular hypothesis, Boyle says, cannot account for the change in the height of the mercury as one ascends the mountain. The problem, of course, was that Linus denied, based on his own experiments, that the mercury changes its height as one ascends. To this Boyle responds that Gassendi vouches for Pascal and says the experiment was repeated five times in varying winds, and indoors and out, and still produced the same results. But he is happy to be able to add that the experiment was also performed by "that known Virtuoso Mr. J. Ball" in Devonshire and by Mr. Richard Towneley in Lancashire, both of whom verified Pascal's general results. (We have no record of Linus's response to Boyle's defense, but he could easily have argued against Gassendi, Ball, and Towneley, as he did against Pascal, that their results were a function of the change in temperature as one ascends the mountain, not of the change in altitude.)

According to Boyle himself, the production of the mathematical expression of the relationship between the pressure and volume of the air, later called Boyle's Law, came about as a result of his attempt to refute Linus's claim that, although the air has some weight and spring, these are insufficient to account for the Torricelli experiment. Linus's funiculus explains those phenomena by its endeavor to hold the mercury up in the tube: the mercury stands at 29½ inches because the funiculus is unable to raise it any higher.

To refute Linus's hypothesis by showing that the mercury can be made to stand higher than 29½ inches, Boyle and his assistants performed a variation of the Torricelli experiment: they took a long glass tube bent into the shape of a J so that the long arm was about 120 inches and the short arm was about 42 inches and sealed at the end. (See Figure 3.) To facilitate measurement, a paper rule, divided into inches, was pasted along the tube. They poured in enough mercury to fill the U at the bottom of the tube and rocked the tube back and forth to insure that the air in the short end was at atmospheric pressure:

> we took care, by frequently inclining the tube, so that the air
> might freely pass from one leg into the other by the sides of the
> mercury (we took, I say, care) that the air at last included in
> the shorter cylinder should be of the same laxity with the rest
> of the air about it. (*W,* I, 156)

Next they poured mercury into the long leg of the tube; Boyle tells us that the weight of this mercury pressed the mercury under it up into the shorter leg, compressing the air there until it took up half the space it had previously possessed. Then Boyle says,

> we cast our eyes upon the longer leg of the glass . . . and we ob-
> served, not without delight and satisfaction, that the quicksilver
> in that longer part of the tube was 29 inches higher than the
> other.

This observation confirmed their hypothesis that "the greater the weight is that leans upon the air, the more forcible is its endeavour of dilatation, and consequently its power of resistance." This was the qualitative hypothesis put forward by Torricelli and others. The quantitative hypothesis Boyle now gives as follows:

> the air in that degree of density and correspondent measure of
> resistance, to which the weight of the incumbent atmosphere

Fig. 3. A J tube. From *W*, I, Plate 1.

Reproduced with permission of the Department of Special Collections,

Stanford University Libraries.

A table of the condenſation of the air.

A	A	B	C	D	E
48	12	00		$29\frac{2}{16}$	$29\frac{2}{16}$
46	$11\frac{1}{2}$	$01\frac{7}{16}$		$30\frac{9}{16}$	$33\frac{6}{16}$
44	11	$02\frac{13}{16}$		$31\frac{15}{16}$	$31\frac{12}{16}$
42	$10\frac{1}{2}$	$04\frac{6}{16}$		$33\frac{8}{16}$	$33\frac{1}{7}$
40	10	$06\frac{3}{16}$		$35\frac{5}{16}$	35- -
38	$9\frac{1}{2}$	$07\frac{14}{16}$		37	$36\frac{15}{19}$
36	9	$10\frac{2}{16}$		$39\frac{5}{16}$	$38\frac{7}{8}$
34	$8\frac{1}{2}$	$12\frac{8}{16}$		$41\frac{10}{16}$	$41\frac{2}{17}$
32	8	$15\frac{1}{16}$	Added to $22\frac{1}{4}$ makes	$44\frac{3}{16}$	$43\frac{11}{16}$
30	$7\frac{1}{2}$	$17\frac{15}{16}$		$47\frac{1}{16}$	$46\frac{3}{5}$
28	7	$21\frac{3}{16}$		$50\frac{5}{16}$	50- -
26	$6\frac{1}{2}$	$25\frac{3}{16}$		$54\frac{5}{16}$	$53\frac{10}{13}$
24	6	$29\frac{11}{16}$		$58\frac{13}{16}$	$58\frac{2}{8}$
23	$5\frac{3}{4}$	$32\frac{3}{16}$		$61\frac{5}{16}$	$60\frac{18}{23}$
22	$5\frac{1}{2}$	$34\frac{15}{16}$		$64\frac{1}{16}$	$63\frac{6}{11}$
21	$5\frac{1}{4}$	$37\frac{15}{16}$		$67\frac{1}{16}$	$66\frac{4}{7}$
20	5	$41\frac{9}{16}$		$70\frac{11}{16}$	70- -
19	$4\frac{3}{4}$	45- -		$74\frac{2}{16}$	$73\frac{11}{19}$
18	$4\frac{1}{2}$	$48\frac{12}{16}$		$77\frac{14}{16}$	$77\frac{2}{3}$
17	$4\frac{1}{4}$	$53\frac{11}{16}$		$82\frac{12}{16}$	$82\frac{4}{7}$
16	4	$58\frac{2}{16}$		$87\frac{14}{16}$	$87\frac{3}{8}$
15	$3\frac{3}{4}$	$63\frac{15}{16}$		$93\frac{1}{16}$	$93\frac{1}{5}$
14	$3\frac{1}{2}$	$71\frac{5}{16}$		$100\frac{7}{16}$	$99\frac{6}{7}$
13	$3\frac{1}{4}$	$78\frac{11}{16}$		$107\frac{13}{16}$	$107\frac{7}{13}$
12	3	$88\frac{7}{16}$		$117\frac{9}{16}$	$116\frac{4}{3}$

AA. The number of equal ſpaces in the ſhorter leg, that contained the ſame parcel of air diverſly extended.

B. The height of the mercurial cylinder in the longer leg, that compreſſed the air into thoſe dimenſions.

C. The height of the mercurial cylinder, that counterbalanced the preſſure of the atmoſphere.

D. The aggregate of the two laſt columns B and C, exhibiting the preſſure ſuſtained by the included air.

E. What that preſſure ſhould be according to the hypotheſis, that ſuppoſes the preſſures and expanſions to be in reciprocal proportion.

Fig. 4. The table of Boyle's measurements leading to the first half of Boyle's Law, i.e., the pressure of the air at or above sea level. From *W*, I, 160.

had brought it, was able to counterbalance and resist the pressure of a mercurial cylinder of about 29 inches, as we are taught by the Torricellian experiment; so here the same air being brought to a degree of density about twice as great as that it had before, obtains a spring twice as strong as formerly. As may appear by its being able to sustain or resist a cylinder of 29 inches in the longer tube, together with the weight of the atmospherical cylinder that leaned upon those 29 inches of mercury; and, as we just now inferred from the Torricellian experiment, was equivalent to them. (*W*, I, 156–157)

In other words, the air in the sealed tip of the tube is dense and resistant enough to resist the pressure of 29 inches of mercury and the air that presses on it. When twice the mercury leans on it, it is twice as dense [i.e., occupies about half the space] and has twice the spring [i.e., can resist the mercury and air above it twice as strongly]. This is the first half of Boyle's Law, describing the behavior of the air at pressures greater than one atmosphere. (See Figure 4, Boyle's original table of measurements.)

In chapters 14 through 18 below, we will discuss whether this experiment refutes Linus's alternative account of suction phenomena. Although we could go on to discuss traditional issues, such as the controversy over whether Boyle deserves the credit for discovering Boyle's Law (other contenders include Richard Towneley, Robert Hooke, and Lord Bouncker) or Robert Hooke's and Isaac Newton's important contributions to the understanding of the law, we will stop here.

What possible part could gender play in the production of Boyle's Law? Turning to the texts by Boyle, Henry Power, Robert Hooke, and others describing original research that led to its production, we find no mention of women or of gender issues. It is true that Boyle discusses women and gender issues, but there is no evidence to date that women participated in the research or made any contribution to the scientific work leading to the recognition that the pressure and volume of the air are in inverse proportion. And even if some woman did make a contribution, what possible relevance could her gender have to the mathematical relationship between the volume and pressure of the air? The irrelevance of external factors such as gender to the technical content of scientific hypotheses, laws, and theories has seemed obvious to many science scholars. An account such as the one we have set out here, showing the intellectual influence of one scientist's work upon that of another, has seemed perfectly adequate to explain this important discovery in the history of science.

To reveal the intersection of gender and Boyle's Law, we must look more carefully at the historical context of Boyle's work, at the activities of women and the responses to their activities, and at the social meaning of such scientific hypotheses as "Nature abhors a vacuum" and "the air has spring and weight," as well as at the technical considerations Boyle advances to support his claims that "Nature is a merely notional thing," "the air has spring and weight," and "the pressure and volume of the air are proportional."

Part Two:
Boyle's Work in Context

Sixteen-year-old Robert Boyle's Grand Tour of Europe was cut short in 1642 by a letter from his father reporting that the Irish had rebelled and that he and three of Robert's brothers were hard pressed to defend his extensive Irish holdings. Robert's brother, Frank, then nineteen, hastened home to Ireland, but Robert returned to Geneva with their tutor; there he learned in 1643 of the death of their father. When Robert Boyle made his way to England in 1644, the country was in the throes of civil war. A brief look at the rich stew of social, political, and religious disputes of the time, including especially the class and gender politics of that turbulent period, will help us to understand his own political and social ideas.

3

Economics, Politics, and Religion
Stuart Conflicts with Parliament

In 1641, the House of Commons enacted legislation that reveals the interests of many of the people who later felt it necessary to overthrow the king. Charles I had often dissolved Parliament when it passed legislation he disapproved of or when it failed to vote him the money he felt he needed. (Because of changes in the economy, bad management, and extravagance, the Stuart kings were always deeply in debt.) The legislation of 1641 included acts that mandated regular meetings of Parliament with or without the king's consent and that prevented the dissolution of Parliament without its consent. The House of Commons also declared illegal all taxes that it did not itself impose, the special courts used to take the property of the nobility and other powerful Englishmen, and the Court of High Commission, which had the power to excommunicate people from the church and to censor the press. Finally, the Commons refused to grant Charles the army he wanted to reconquer Ireland, which had just rebelled against English rule. Charles tried to arrest the leaders of the Commons who opposed him; when he failed, he left London, and the English Civil War began.

These Parliamentary actions reveal some of the issues of struggle between the English monarchs and the court party, on one side, and the rising capitalist class represented by Parliament—merchants, traders, investors—on the other. These struggles were not new; throughout the seventeenth century, the Stuart kings battled for a return to an absolute monarchy, while the "middling sort of people" struggled against absolute monarchy and for absolute control of their own property, for a say in how their country was governed, and for a measure of freedom of religion. In 1628, Parliament had presented to Charles I the Petition of Right, claiming, among other things, that taxation without Parliamentary consent and imprisonment without cause were illegal. These claims were made in part because, in return for the hand of the Catholic Henrietta Maria of France, Charles had agreed to allow Catholics to practice their religion in England and to help Louis XIII put down the rebellious Protestants of La Rochelle. Parliament strongly objected and was dissolved without voting Charles the money he wanted and needed. But the king managed by collecting "tonnage and poundage" (that is, customs dues) without Parliament's consent and by a forced loan (a tax in all but name). Eighty gentlemen who refused to pay were imprisoned without being charged, while poorer men were pressed into the armies raised to fight France and Austria.

Not all of Charles's attempts to raise money were so legally dubious. The Stuarts sold monopolies on hundreds of products, from pins and gloves, glass and iron to salt and lobsters, and these sales brought great revenues. But with monopolies on desirable goods, prices went up, contributing to the disappointment or financial distress of those directly affected and to inflation, which hurt the crown as well as everyone else. Those who lost their investments when the monopolies went to their rivals and those who were unable to enter a lucrative or promising industry were naturally disaffected. And those holding a monopoly could not, even then, be sure of financial success, since the Stuarts often sold monopolies twice to different people.[1] Ultimately, the greatest beneficiaries were the courtiers who, in return for using their influence over the granting of monopolies, wrested a percentage of profits from their holders.

The kings of England also had a right of wardship. When a landowner who received his lands from the king died and the heir was a minor, the king had the right to become the guardian of the ward and to run his estates and profit from them until the ward reached majority (Hill, 1961, 49). And finally, ship money (a tax to pay for shipbuilding) provided perhaps the greatest source of new revenue for the government. Despite the Petition of Right, Charles taxed the inland towns of England without the consent of Parliament by collecting ship money from them even though they were not port towns.

When Henry VIII had become the head of the church in England, the church had become the state church, the Church of England; thus, there was in principle no separation between church and state. Many political issues which modern English speakers are accustomed to think of as matters of secular politics were disputed and understood in seventeenth-century England in religious terms. One of the greatest of these issues was the issue of sovereignty: by what right did the king claim power (divine right), by what right Parliament (the will of the people)?

Before William Laud, who became primate of the Church of England in 1633, began his efforts to enforce conformity of both belief and form of worship, local churches practiced a variety of forms of worship and included people whose beliefs were in fact quite diverse; the Church of England was, in a sense, comprehensive. Using the Court of High Commission and a system of visitations, Laud sought to impose conformity, particularly upon rising Puritan sentiment. The High Commission not only maintained Laud's censorship of the press but also censored and punished lecturers and conventicles of sects as well as congregations and the clergy. It was particularly hated for requiring that prisoners incriminate themselves under oath. As clergy who fell afoul of Laud's visiting deputies or his inquiring bishops were turned out of their livings or frightened into silence, town and city corporations hired "lecturers" who spoke in the parish church after the service was read and gentlemen employed private chaplains to serve this function. Laud put an end to both these practices.

Those souls who found the imposed ritual—which they were required by law to endure each Sunday—unedifying and the censorship of their beliefs stifling either set off for the American colonies, hoping to practice religion according to their own lights, or they joined conventicles, groups that met in secret to pray, study the scriptures, and hear preachers more to their taste. But conventicles were the special object of Laud's zealous persecution and those who were caught found themselves tried and convicted of dissent. Trevelyan notes that those who frequented the conventicles were "men [sic] of mean station" and that records of the trials of dissenters show that they were trials of poor people.[2]

Ritual was only the most palpable of a number of issues at stake in the seventeenth-century dispute over religion in England (as well as in Scotland and Ireland). Many feared that the husband (Charles I) and then the sons (Charles II and James II) of Henrietta Maria wished to restore England to the Church of Rome. Any move in that direction fed their fears, from the failure of the Stuarts to fight for Protestantism on the continent to Henrietta Maria's favoring of papists at court

to Laud's attempts to impose conformity on the church. High on the
list of worrying moves was the claim of Anglican bishops to be called
to office by divine right; they claimed apostolic succession just as Roman bishops did. Both James I and Charles I defended the bishops
against their critics, and the bishops upheld the divine right of kings
in return. The mutual support of the monarchy and the church was
essential; James's aphorism "No bishop, no king" was echoed by the
divines in Charles's court who argued that "those who were eager to
cast mitres and copes under foot were equally anxious to throw down
crowns and sceptres," and Laud himself argued that the unity of the
polity depended upon unity in the church. On the other side, Puritans
argued that popery and tyranny went hand in hand (Davies, *Stuarts*,
69 and 71).

4

Civil War Approaches

Class politics became clearer as the Civil War approached. Royalists feared that the lower classes wanted social revolution (and as we shall see, many people certainly agitated for radical economic and political change). And many of the "middling sort" felt the need for political and economic reform. In chapter 6 we will take a brief look at the Boyle family's complex class politics.

In 1637 Charles I and Laud ordered that conformity be imposed upon the Presbyterian Church of Scotland and that the Anglican Book of Common Prayer be used in all services; Charles also tried to re-annex church and crown lands taken over by Scottish nobles at the time of the Reformation and to rule Scotland without the counsel of these nobles. The Scots responded by setting up a select committee to rule Scotland and by swearing to a covenant according to which they refused the Laudian innovations to the death (but promised to defend the king and the true religion). Charles capitulated, but when the Scots elected a general assembly (the ruling body of the Presbyterian Church) whose membership displeased him, he declared it dissolved. It remained sitting, however, and abolished not only the use of the prayer book but episcopacy itself. The first Bishops' War followed.

Charles had no standing army; his only options were train-bands, which were poorly trained and poorly armed local militia, or impressed peasants. He called upon the train-bands of the northern counties and the nobles of England, most of whom were disaffected from him. For example, Boyle's father, Richard, the Earl of Cork, then residing in Dorset at Stalbridge, contributed £3000 to help his eldest son, Lord Dungarvan, raise and arm "a gallant troop of horse for the King's service," though he thought Dungarvan's undertaking to fight was "an unadvised engagement." Nevertheless, he allowed two younger sons, the Lords Kinalmeaky and Broghill, "to accompany him in that service" (*W*, I, xix). The king's preparations also included sending Robert Boyle's brother-in-law, Lord Barrymore, to Ireland with letters urging Strafford to raise a thousand foot for the army.

Nevertheless the king's forces were no match for the Scottish veterans of the armies of Sweden and Holland, and after brief negotiations, Charles agreed that Scotland's general assembly should determine the course of the Scottish church and its parliament the course of the state. But within the year he began preparations for a second war.

In 1640, Strafford, recently returned from Ireland, persuaded the king to summon Parliament to raise money for an army, but the English parliament was much more concerned to recover the privileges it had lost to Charles's personal rule than to advance him the subsidies which would allow him to extend that rule. The king dissolved this Short Parliament, so called because it sat so briefly, and riots ensued around England.

The Convocation of Bishops passed canons designed to reinforce the existing church and the divine right of kings, and Strafford persuaded the king to raise funds by a forced loan from the City of London, as well as more ship money and coat and conduct money for the train-bands. Four London aldermen who refused the loan were imprisoned, but meanwhile, press-gangs in the south of England raised a (mutinous) army which was moved toward Scotland.

Again Charles was forced to capitulate to Scottish demands, which now included an indemnity and £850 per day to keep the Scottish soldiers in Northumberland and Durham until the indemnity was paid. For weeks Charles resisted calling a Parliament, but, under threat from the Scots, he had no other choice, and the Long Parliament sat in November 1640.

Almost immediately Parliament began to consider how Strafford might be brought down and, in turn, Strafford persuaded the king to arrest its leaders for treason (plotting with the Scots). Thereupon, the House of Commons hastily drew up an impeachment of Strafford and he was imprisoned in the Tower. When it became clear that the House of Lords would not after all impeach him, the Commons passed a bill of attainder for his death and sent it to the Lords, who accepted it, influenced by Charles's refusal to disband the Irish army and by rumors of the threat of force against Parliament. When a mob composed of both middle-class citizens and the rabble surrounded Whitehall and threatened the king and queen, Charles signed the bill and Strafford duly lost his head. Other ministers were impeached and Laud was imprisoned in the Tower.

In the summer of 1641 Charles left England to visit Scotland over the strong objections of Parliament, which rightly feared that he sought the aid of the Scottish army to bring his unruly English subjects back to order. When he left, Parliament illegally issued temporary ordinances for the security of military stores and returned to its great debate over the nature of the state church. Members were uncertain of what they wanted, but they knew what they did not want: bishops moving toward Catholicism and a king toward absolute monarchy. In particular, two possibilities were untenable to them: continuing the current episcopalian system with bishops appointed by Charles and granting the freedom to worship as one saw fit.

In the autumn of 1641, before Charles returned from Scotland, the Irish rebellion broke out. Many in England understood the rebellion to be a Catholic attempt to massacre the Protestants there, leading members of Parliament to forget their divisions over religion and to join in voting Charles the army he had so long desired. They differed, however, over who should lead it: Pym and his party voted for officers named by Parliament, but the episcopalians carried the day in support of the king. In a desperate appeal to the people, Pym's party pressed for the passage of the "Grand Remonstrance," documenting years of misgovernment in the state and calling for a "general synod" to decide the matter of church government. This document was presented to the king and to the people. When Parliament at last introduced a bill to nominate a lord general who could raise an army, Charles gathered his first group of "cavaliers" and replaced the Puritan Balfour with their leader as lieutenant of the Tower. And he tried, illegally, to impeach five leaders of his opposition in Parliament. These men fled to the protection of the City of London, which appointed a major general to command the London train-bands for the support of Parliament. There was an outpouring of support for Parliament against the king, and Charles fled London.

But the following months saw half the nation rally to the king, driven in part by fear of the sectaries, for "sober puritanism was being replaced by enthusiastic Brownism."[1] Sectarian sermons calling for toleration and for new ways threatened to undermine not only the church but the established social order. By September 1642, the class politics had become clear; royalists feared that the lower classes wanted social revolution, crying that the upper classes "have been our masters a long time and now we may chance to master them, and now they [the lower classes] know their strength it shall goe hard but they will use it." And the Independent Edmund Ludlow stated that most of the peers and great landowners followed the king to protect their own interests: "Many of the nobility and gentry were contented to serve his [Charles's] arbitrary designs, if they might have leave to insult over such as were of a lower order."[2]

5

The Intersection of Class
and Gender Politics

In this chapter we will see the beginnings of a radical threat to the status quo that included both class and gender threats: the radicals threatened a redistribution of decision-making authority as well as a redistribution of wealth. And the activities of women during the Civil War posed a threat to decision-making authority not just in the family or private sphere, but also in the public sphere. This threat began within the sectarian community, i.e., the "gathered" churches, but extended to the streets and ultimately to the state.

The culmination of the English Civil War is sometimes referred to as the Puritan Revolution, because the fight was religious as well as political and economic and because many of the political and economic issues were debated in religious terms. The very name "Puritan Revolution" tells who won: the Puritans defeated the royalists, but they also defeated the radicals.

The radicals were included among several groups of "masterless men" (and women) composed, first, of rogues, vagabonds, and beggars; second, of casual laborers in London, such as dock workers, watermen, building laborers, and journeymen, as well as fishwomen— all those people who made up "the mob," as it was called; third, of the rural poor, including cottagers and squatters on commons and in wastes and forests; and, finally, the Protestant sectaries: townspeople, often immigrants, who were small craftsmen, apprentices, and "serious-minded laborious men" who rejected the state church. Instead of the hierarchical society supported by the doctrines of the Church of England, the sectaries preferred a more democratic society, supported by their belief that God is in all his "saints" (as they called themselves), so they do not need priests of the established church to mediate between them and him. In chapter 11 we will discuss the metaphysics underpinning sectarian religious beliefs and political aspirations. Here we need only note that for sectaries, each individual

53

has access to God; each is responsible to God for his or her own soul.[1] We are not surprised, then, to learn that political and religious groups often overlapped. For example, Brailsford describes the Levellers as a "middling sort of people," including "craftsmen, cobblers, weavers, printers and lead miners, together with some well-to-do tradesmen," some journeymen, and on occasion a professional man. Some attended Independent churches, but large numbers of them, if not the majority, were sectaries.[2] On the other hand, many sectaries shared "leveling" views but did not belong to any political organization.

Sectarian emphasis upon the individual soul had important implications for sectarian women. The seventeenth century saw the development of the ideal woman as a bourgeoise who was to marry and to stay at home minding the house; while married, she was to own no property. She had no voice in the church or state. Puritan marriage manuals continually reinforced the view that "the man when he loveth should remember his superiority"[3] and William Gouge, in his popular manual *Of Domesticall Duties* of 1622 and 1634, flatly declared that "the extent of wive's subjection doth stretch itself very far, even to all things."[4]

But the rise of sectarianism, with its view that God is in everything and everyone, threatened the sexual status quo. The Leveller John Lilburne remarked that "Every particular and individual man and woman that ever breathed in the world since [Adam and Eve] are and were by nature all equal and alike in power, dignity, authority and majesty, none of them having (by nature) any authority dominion or magisterial power, one over . . . another."[5] Thus, all members of sectarian congregations, including women, debated, voted, prophesied, and even preached. Too, since the sectaries believed that the regenerate must separate from the ungodly, sectarian women were often allowed or encouraged to divorce or separate from their unregenerate husbands.

For the Quakers, the principle that the spirit of life is in all creatures found expression in the doctrine of the inner light. "An immediate effect of the experience of the Inner Light was that the recipient recognized the God-within and began immediately to operate out of the assumption that she/he now possessed a spark of divine life."[6] In London, the spirit led many sectarian women to preach and to travel as preachers and missionaries to the university towns as well as to Ireland and America; Quaker women also traveled to Italy and to the eastern Mediterranean to spread their message. Thus, for example, "In 1653 Mary Fisher and Elizabeth Williams went to Cambridge, where they 'discoursed about the things of God' with the young theologians and preached publicly at the gate of Sidney College. Their behavior so upset the mayor of Cambridge that he ordered them to be taken to Market Cross, where they became the first of the Friends to be publicly scourged" (Huber, "Woman," 163). Elizabeth Fletcher, aged 17, and her friend Elizabeth Leavens received about the same treatment at Oxford in 1654.

At various times during the Civil War and the years before the restoration of the monarchy, sectaries, Levellers, and others called for a number of more or less revolutionary reforms. They wanted an end to enclosure, the practice of enclosing and developing waste land; such development required that the people living on the land, who had sometimes been there for generations though without any explicit property rights in it, be displaced and rendered homeless. They also objected to rent racking, sharp increases in the rent owed by tenant farmers to landlords. Like enclosure, fen drainage yielded more land for cultivation and was justified by its forcing squatters "to quit idleness and betake themselves to . . . manufactures."[7] The people whose homes and livelihoods were lost to fen drainage and enclosures fought back by all available means. As early as 1603 women led a revolt against the drainage of Deeping Fen in Lincolnshire, and others participated in the destruction of enclosures in Braydon Forest in the 1630s, at Buckden in 1641, at York in 1642, and in other places.

Unfortunately, these economic struggles divided women by class. In Yorkshire, the upper-class Margaret Eure complained about lower-class women in a letter to Sir Ralph Verney,

> I wish you would take heed of women for this very vermin
> have pulled down an enclosure, which some of them were put
> in prison for it by the justices[;] they had their pipe [of to-
> bacco] to go before them and their ale and cakes to make
> themselves merry when they had done their feats of activity.[8]

Though they did not always share the same political or economic interests, women on all sides acted on the issues that concerned them. For example, the wife of one of the members of the House of Commons (Captain Venn, who sat for the City of London) "had with great industry solicited many people to go down with their arms to Westminster" to put pressure on the House to act on the Grand Remonstrance. And in January 1641/2 the Lord Mayor of London accompanied Charles when he came to arrest five members of Parliament, and according to one report, when the king departed, "the citizens' wives fell upon the Lord Mayor, and pulled his chain from his neck, and called him traitor to the liberties of it, and had like to have torn him and the Recorder in pieces" (quoted in Higgins, "Reactions," 184).

Soon thereafter, "a company of women" moved by deteriorating economic and political conditions petitioned the House of Lords to end its recalcitrance in settling its differences with the Commons (which, according to the women, "did what in them lay to relieve them, and redresse their grievances"). Specifically, the petitioners were concerned about "the great decay of trading" and that "Religion may be established, and present aid and assistance transported into Ireland for the reliefe of the distressed Protestants." The next day women crowded into Palace Yard—the open space that lay outside the House of Lords—declaring that "We had rather bring our children and leave them at the Lords' door, than have them starve at home." They may have been among the four hundred women who returned with a second petition "desiring an answer to their petition delivered the day before." These women presented their petition to the Duke of Lennox, who answered, according to one source,

> "Away with these women, wee were best to have a parliament of Women." Whereupon some of the Women interrupting his passage, catched hold of his staffe, humbly desiring him to receive their petition, upon which the Duke being moved, offered to draw back his staffe, but they holding it so fast between them, it was broken, whereupon the said Duke was enforced to send for another staffe; after which they delivered their petition to the Lord Savage. (Quoted in Higgins, "Reactions," 185–186)

These women managed to get twelve of their number called into the House of Lords to explain their grievances. They were more successful than their sisters who petitioned the House of Commons at the same time, for despite the threat that "where there is One Woman

now here, there would be Five hundred To-morrow; and that it was
as good for them to die here, as at home," the House instructed
Sergeant-Major General Skipon to inform them that the House was
"now in Consideration of Matters of great Consequence; and will
hereafter give such Directions as the Occasion shall require" (quoted
in Higgins, "Reactions," 186–187).

Women petitioners were very persistent, for three days later "a
company of Women" returned with a second petition demanding an
answer to their first petition against the bishops sitting in the House
of Lords. These women declared that "great Danger and Fear do still
attend us, and will, as long as Popish Lords and superstitious Bishops
are suffered to have their Voice in the House of Peers." Moreover,
"that accursed and abominable Idol of the Mass [is] suffered in the
Kingdom, and that Arch-enemy of our Prosperity and Reformation
[i.e., Archbishop Laud] lieth in the Tower yet not receiving his de-
served Punishment" (quoted in Higgins, "Reactions," 187). Despite
their apparent success (the Lords agreed the next day to exclude bishops), persistence was necessary because women had no legal standing
to petition. This problem led these petitioners, such as the gentlewoman and brewer's wife Mrs. Anne Stagg and "the many others
with her of like Rank and Quality," to defend their right to petition.
Their argument was based first on the religious equality with men
stressed in their Puritan and sectarian congregations and teachings:

> It may be thought strange . . . to shew ourselves by Way of Peti-
> tion, . . . but [considering] the Right and Interest we have in
> the common and publick Cause of the Church, it will . . . be
> found a Duty commanded and required.

But the women also claimed an equal interest in the commonwealth:

> Women are Sharers in the common Calamities that accompany
> . . . Church and Commonwealth when Oppression is exercised
> over the Church and Kingdom wherein they live, and an un-
> limited Power hath been given to the Prelates, to exercise
> Authority over the Consciences of Women, as well as Men;
> witness Newgate, Smithfield, and other Places of Persecution,
> wherein Women, as well as Men, have felt the Smart of their
> Fury. (Quoted in Higgins, "Reactions," 215, 216)

We must note that political activity in the form of petitioning was not limited to women of the Parliamentary party. Royalist women, too, pressed for attention to their economic needs and political demands. A few days after the petition against the bishops, the Lords were again petitioned by "divers of her Majesties servant wives" to persuade the Queen and Lady Mary not to leave for Holland. Their reasons may have been economic; a tract describing a petition delivered on the same day for the same purpose mentions that "an utter cessation and decay of all our trading" would follow the Queen's departure, since

> your Petitioners, their Husbands, their Children, and their
> Familes . . . have lived in plentifull and good fashion, by the
> exercise of severall Trades and venting of divers workes . . . All
> depending wholly for the sale of their commodities (which is
> the maintenance and very existence and being of themselves,
> their husbands and families) upon the splendor and glory of
> the English Court, and principally upon that of the Queenes
> Majesty. (Quoted in Higgins, "Reactions," 188)

Women's demonstrations at Parliament began to be more violent as the Civil War deepened and economic and social conditions became more chaotic. In August 1643, both royalists and those supporting Parliament petitioned for and against a peace proposal put forward by the House of Lords. News sheets claimed that royalist men instigated the demonstrations for a peace on royalist terms, but pointed out that peace *"was to the women a pleasing thing, and thereupon some out of an earnest desire of Peace,* others out of the designe, came on Tuesday to Westminster . . . and cryed for Peace." The women were described as "oyster wives, and other dirty and tattered sluts" or as "the wives of substantial citizens" depending upon the politics of the observer. On August 8, the demonstrations began to turn nasty; "the women 'so filled the staires' of the House of Commons 'that no man could passe up or downe, whereupon a man upon the top of the staires, drew his sword, and with the flat side stroak some of them upon the heads, which so affrighted them, that they presently made way and ran downe, and thereupon the Trained band that then waited, came and made a Court of guard upon the staires and so kept them off from further troubling the House."

But on August 9, the number of women swelled to "five or six thousand" and

> there was such a fearfull Tumult and uprore raised by women
> about the Parliament House; as was never recorded by any His-
> tories either ancient or moderne, which by eye witnesses is
> thus related. After the Trained Band was come into the Old Pal-
> ace yard, and had set their Sentinell at the usuall places accord-
> ing to the custome, about an hundred women with white Rib-
> bands in their hats pressed to make way through them, which
> the Sentinells opposing, more women come to second them,
> fell upon them and beate them away, and by violence made
> their way into the yard; then more women striving to land at
> the Parliament staires, were kept of[f] there by the Sentinells,
> but they landing a little higher came in upon the West side of
> the yard, and then all of them cried out mainely, we will have
> Peace presently and our King.

These women petitioned the Commons for "the just Prerogatives and
Priviledges of King and Parliament"; that "the true Liberties and
Properties of the Subject" be restored; that "some speedie Course
maie be taken for ye settlemt of ye true Reformed Protestant Religion
. . . and ye renovation of Trade."

The women were told that the House of Commons supported peace
and were asked to go home, but they "were so farre from being satisfied that Sir John Hepsley and the rest received such course usage
from them, that they desired no more of such imployment." D'Ewes
reported that the women fell "upon all that have short haire and pull
them both Ministers, Souldiers and others"; moreover, "a Minister
passing through the yard, they laid hands upon him, cried out a
Roundhead, and tore his cloak and band." Violence then broke out in
earnest: "From words they fell to blowes"; for during the afternoon,
the women gained the top of the stairs into the House and kept the
train-band out, pushing them "downe by the head and shoulders."
They had the members cooped up inside the house, resisting the
train-band's advice to come down; nor were they persuaded by the
powder shot at them. Indeed, they threw pieces of brick "very fast at
the trained Bands, and disarmed some of them . . . these women were
not any whit scared or ashamed of their incivilities, but cryed out so

much the more . . . , 'Give us these Traytors that are against peace, that we may teare them in pieces, Give us Pym.'" The House guards were apparently "loath to offer violence to women," but when the women attacked a troop of horse, the troopers

> drew their Swords, and laid on some of them with their Swords flatwayes for a good space, which [the women] regarded not, but enclosed them; upon this they then cut them on the face and hands, and one woman lost her nose, whom they say is since dead, as soon as the rest of the women saw blood once drawne, they ran away from the Parliament House, and dispersed themselves in smaller numbers, into the Churchyards, Pallace, and other places; and about an houre after the House was up, a Troope of horse came, and cudgelled such as staid, with their Kanes, and dispersed them.

According to one source the women declared that they would return the next day "with greater strengthe and would have swords and guns likewise" and threatened to "demolishe all the workes [defensive fortifications] aboute the towne" (Higgins, "Reactions," 190–196).

Although to us the suggestion of the women petitioners that as women they shared with men in the "common Calamities" seems modest enough, we must not forget that the act of petitioning itself, let alone demonstrating at Parliament, challenged the current status of women in the state. As women, they had no right to petition, but by petitioning they implicitly claimed that right. And certainly their willingness to resort to physical threats indicates their determination to press both their explicit and their implicit claims. Moreover, the exclamation "wee were best to have a parliament of Women" is ironic, but expresses a recognition that the women petitioners were in fact threatening the current decision-making arrangements of the state. *The Parliament Scout* drove home the moral for gender politics: "Thus we see, to permit absurdities, is the way to increase them; Tumults are dangerous, swords in women's hands doe desperate things; this is begotten in the distractions of Civill War" (quoted in Higgins, "Reactions," 198).

6

The Boyle Family's Religious and Class Politics

By the time Boyle arrived in England in mid-1644, Charles had fought and won several battles against Parliamentary forces, but in July 1644 he lost the Battle of Marston Moor when the Scottish army joined the Northern Army and Cromwell's Eastern Association. Cromwell's troops were a marked contrast to the other Parliamentary forces as well as to the royalists. The latter were composed of ill-paid, impressed foot soldiers and cavalry led by local nobility whose claim to leadership was based on ancient custom; Cromwell, however, drew on men of strong Puritan beliefs whose inclinations followed his insistence both upon strict military and moral discipline and upon more regular pay. When he later persuaded the Parliament to raise its own "New Model Army," he founded it on the same principles. Sectaries and the "the mob" were among the foot soldiers and Cromwell insisted that no religious test be applied to officers, leaving the ranks open to Independents whose strong feeling against royalists and a state church made them fierce soldiers. Officers were commissioned based on ability and merit rather than social rank, and in 1645 this New Model Army, despite its untried new recruits, defeated Charles's army at Naseby. Thereafter, Parliamentary forces laid siege to and overcame the royalist strongholds and in May of 1646 Charles fled to the Scots.

The political loyalties of the Boyle family were complex. Although Robert Boyle's father, Richard, Earl of Cork, was English, his wealth was Irish; his titles were given him by the English king, but they were Irish titles. And the Irish rebellion—which broke out at about the same time as the English Civil War and dragged on even longer—was a very complicating factor, indeed. Canny explains that the Civil War forced Irish planters like the Earl of Cork to look for support from whoever appeared to be in control in England. For the old Earl, the king represented authority, but when the balance of power in England shifted to Parliament, then to the Lord Protector, and finally to Charles II,

> Cork's successors in Ireland, but especially Roger, Baron Broghill, who was principally concerned with the defence and resumption of the Munster lands [the center of the Boyles' wealth], gave allegiance to whichever force happened to represent stability in England and hence security for the planters in Ireland.[1]

On the other hand, Robert's eldest brother, Richard, held substantial property in England and took the royalist side in the Civil War. And when Boyle arrived in London in 1644, he intended to join the king's army, but instead lived for several months with his "beloved sister," Lady Katherine Ranelagh. Birch tells us that "a sister-in-law of lady Ranelagh, who was with them in the house, and was wife of one of the principal members of the then house of commons, brought him into the acquaintance and friendship of some great men of that party, which was then growing, and soon after victorious; by whose means he got early protection for his English and Irish estates" (W, I, xxvii). In fact, Lady Ranelagh herself was closely connected with the Parliamentary party and with the great poet of the Puritan revolution, John Milton, who tutored her nephew and who later wrote to a friend that "she stood 'in place of kith and kin' to him in his blindness and solitude."[2] These Parliamentary connections made it possible for Boyle to establish his title to Stalbridge, the estate in Dorset left him by his father.

After taking up residence at Stalbridge in 1645, Boyle frequently traveled to London, where he stayed with Lady Ranelagh and was well apprised of the activities of Parliament and the tumults of the capital. In 1646, for example, he wrote from London to his old tutor describing events in England and indicating his sympathy with the Parliamentary forces:

> In England there is not one malignant garrison untaken, and in Wales but two or three rocky places hold out for the King, and these too so inconsiderable, that they more advantage their enemies forces, by keeping them from idleness, than they are able to prejudice them by their opposition. The Scots being

now to quit the kingdom, the parliament had compounded
with them for all their arrears, upon whose payment they are
to deliver up their garrisons, and retire into their own country.
. . . His majesty is still at Newcastle, both discontenting and dis-
contented; and the Scots will now, upon their departure out of
England, be forced some way or other to dispose of his person,
which the houses have here voted to remain at the disposition
of both houses of parliament. (*W*, I, xxx–xxxi)

During the Civil War, both the sectaries and "the mob" were called
upon to aid in the revolution by those who opposed the king; for example, they were, as we have seen, among the foot soldiers in the
New Model Army used by Cromwell to defeat Charles I. Once called
upon to act, however, these people of the "lower orders" began to
think and to speak for themselves and discovered that their political and economic interests conflicted not only with those of the king
and his party, but also with those of the "middling sort" who were
running the revolution. The differences between the social vision of
Cromwell and other "grandees" on the one side and the rank and
file, the sectaries and the Levellers in the army, on the other were
made clear in 1647 in an extraordinary series of events that began with Parliament's foolish decision to disband the Parliamentary
army without paying the wages of the soldiers—then seriously in
arrears—and without responding to their need for indemnity for actions of war.

By the time Parliament ordered the New Model Army to disband,
it had established Presbyterianism as the state religion and passed bills
to suppress the sectaries: "Unitarians and free-thinking heretics could
be put to death, and Baptists and other sectaries imprisoned for life,
solely on account of their dogmatic opinions," and lay preachers were
subject to strict punishment.[3] In this matter, the opinions of Robert
Boyle were complex. In 1646 he was alarmed by the sectaries, but
inclined to toleration:

The presbyterian government is at last settled (though I scarce
think it will prove long lived) after the great opposition of
many and to their no less dislike; though it seemed very high
time unto others, that some established and strict discipline
should put a restraint upon the spreading impostures of the sec-

taries, which have made this distracted city their general rendezvous, which entertains at present no less than 200 several opinions in point of religion, some digged out of those graves where the condemning decrees of primitive councils had long since buried them; others newly fashioned in the forge of their own brains; but the most being new editions of old errors, vented with some honourable title and modern disguisements; so that certainly if the truth be any where to be found, it is here sought so many several ways, that one or other must needs light upon it. But others, that justly pretend to a greater moderation, suspect, that our dotage upon our own opinions makes us mistake many for impostures, that are but glimpses and manifestations of obscure or formerly concealed truths, or at least our own pride or self-love makes us aggravate very venial errors into dangerous and damnable heresies. The parliament is now upon an ordinance for the punishment of many of these supposed errors; but since their belief of their contrary truths is confessedly a work of divine revelation, why a man should be hanged, because it has not yet pleased God to give him his spirit, I confess, I am yet to understand. Certainly to think by a halter to let new light into the understanding, or by the tortures of the body to heal the errors of the mind, seems to me like the applying a plaster to the heel, to cure a wound in the head; which doth not work upon the seat of the disease. (*W,* I, xxxii–xxxiii)

Boyle's own views at this time were influenced by a circle of friends including Samuel Hartlib and John Dury, whom Boyle described as "men of so capacious and searching spirits, that school-philosophy is but the lowest region of their knowledge. . . . persons that endeavor to put narrow-mindedness out of countenance, by the practice of so extensive a charity, that it reaches unto every thing called man" (*W,* I, xxxiv–xxxv). Hartlib and the others hoped that Parliament would help to reform England along the lines set out in a tract written by Gabriel Plattes, *A Description of the Famous Kingdom of Macaria.* Here we find a utopia in which the king, like other landowners, follows the agricultural advice of the "college of experience," and is able to live of his own and so need not overburden his subjects with taxes. The kingdom is prosperous and strong, and as a result of its strength, it is peaceful. The peace is further strengthened by the state church, for "[i]f any divine shall publish a new opinion to the common people" before it has won the agreement of the great council, "he shall be accounted a disturber of the peace, and shall suffer death for it."[4] Boyle and the Hartlib circle felt that war resulted when men strayed from the true religion into sin; the solution to war and civil strife, then, was to return men to reasonable religion. Boyle agreed with Dury and Hartlib that the most reasonable settlement could be found by requiring only those tenets which reasonable Christian men could agree to and leaving the rest to individual conscience. No one would be forced against his conscience into strict conformity with inessential doctrine, but everyone must belong to the same state church; people would not have complete freedom of religious opinion, for that leads to dissension and atheism (in seventeenth-century terms, any belief or conduct regarded as un-Christian). A reasonable settlement was understood to be both religious and political. The state church would be tolerant within limits and the state would ideally resemble Plattes's Macaria, characterized by a monarchy balanced by a strong landed class; social hierarchy, gender hierarchy, and social differences would, of course, carry with them economic differences.

Boyle wanted a moderate church settlement, but by 1647 this did not include "the worst part" of the sectaries:

As for our upstart sectaries (mushrooms of the last night's
springing up) the worst part of them . . . smitten at the root
with the worm of their irrationality, will be as sudden in their
decay, as they were hasty in their growth; and indeed perhaps
the safest way to destroy them is rather to let them die, than attempt to kill them. (W, I, xl)

He was certain that they needed to be destroyed, since they were proving a danger to the polity. That danger resulted not from their religious views alone but from political views that in part arose from the religious ones, and in part comported well with them. Let us examine some of these political views.

7
More Class and Gender Politics

In this chapter, we take a brief look at the extraordinary series of radical debates and activities that preceded the restoration of the monarchy under Charles II. The "lower orders" in the army and the streets proposed many revolutionary social arrangements and women played an important part, writing petitions, gathering signatures, and demonstrating at Parliament. Responses to these activities reveal the perceived threat not only to the gender hierarchy but to the system of class hierarchy itself.

When, in 1647, the army refused to disband, Parliament made plans to use garrisons of cavaliers and the London city militia to force it to. Cromwell, as a member of Parliament, had tried to persuade the House of Commons to pay the soldiers, but he took up leadership of the New Model Army at the last moment. Two months of negotiations failed to reconcile Parliament and the army and in August 1647, the army occupied London without bloodshed. Its negotiating positions were to be worked out by a council, an extraordinary democratic experiment consisting of officers and representatives of each regiment, later called "agitators." The council produced several remarkable documents setting out alternative revolutionary social arrangements. The documents included the Solemn Engagement (setting out their grievances), the Declaration of the Army, the Heads of the Proposals, and an Agreement of the People. The Heads of the Proposals suggested legislation designed to provide for a strong Parliament:

> parliament should set a date for its own termination; henceforth there should be biennial parliaments . . . enjoying control over the army and navy for ten years; no royalist should be permitted to hold any office in the state for five years or to be elected as a member until the second biennial parliament was over; there should be a council of state, composed of persons . . . who should conduct all foreign negotiations but should require

the consent of parliament to make war or peace; parliament should nominate great officers of state . . . ; all coercive power should be taken away from bishops and other ecclesiastical officers; all acts should be repealed enjoining the use of the Book of Common Prayer [and conventicles]; no one should be compelled to take the [Presbyterian] covenant; and royalists should be allowed to compound on easy terms for their delinquency.[1]

Further reforms should include

a vindication of the right of petition, the prompt lifting of the excise from commodities necessary to the poor, the abolition of all monopolies . . . , "some remedy for the present unequal, troublesome and contentious way of ministers' maintenance by tithes", the simplification and cheapening of legal procedure, a reform of the present practice of imprisonment for debt and an endorsement of the principle that no man may be compelled to incriminate himself.[2]

H. N. Brailsford notes that this document adopts nearly every item set out in the Levellers' political program (Brailsford, *Levellers,* 245).

The Leveller leader, John Lilburne, had argued that the Civil War put English society back into a state of nature, and the Levellers believed that a social contract was the proper way to reconstitute society. They thought of the social contract as analogous to the Independent and Anabaptist "gathered" churches, which were marked by voluntary association with "a closely knit community and the signing of a covenant with each other for mutual aid and discipline" (Brailsford, *Levellers,* 259). The terms of the social contract were not mere concepts to Lilburne; he had refused to incriminate himself by answering the questions of the Star Chamber, the Lord Keeper, and the Lord Privy Seal and had been sentenced to pay a £500 fine and to be publicly flogged and pilloried. He was imprisoned for three years, until Laud's reign ended and the Long Parliament resolved that his punishment was illegal. Though he was no longer in the army when the Putney debates took place, his ideas were.

The debates at Putney church between the radicals and the conservatives in the army reveal the extent to which Leveller ideals met with the strong response of the rank and file and of many officers. All manner of pamphlets were circulating in the army; the men heard sermons and participated in discussions of Leveller ideas such as those mentioned above. But the most controversial demand made by the radicals was for manhood suffrage: "We judge," they declared, "that all inhabitants that have not lost their birthright should have an equal voice in elections." In support of this claim, Colonel Rainsborough, speaking for the radicals, declared

> that the poorest he that is in England hath a life to live, as the greatest he; and therefore, truly, Sir, I think it's clear that every man that is to live under a government ought first by his own consent to put himself under that government; and I do think that the poorest man in England is not at all bound in a strict sense to that government that he hath not had a voice to put himself under. . . .
>
> I do not find anything in the law of God that a lord shall choose twenty burgesses and a gentleman but two, or a poor man shall choose none; I find no such thing in the law of nature, nor in the law of nations. (Both quotations in Brailsford, *Levellers*, 274–275)

This demand was immediately recognized as a threat to private property. Speaking for Cromwell and the conservatives, General Ireton declared that "If you admit any man that hath a breath and being . . . this will destroy property. . . . Why may not those men vote against all property?" (quoted in Brailsford, *Levellers*, 276). Cromwell and the leaders of the revolution wanted suffrage only for men of property (usually understood to mean landed property), and they recognized that men of no property outnumbered them and might vote to redistribute wealth.

They were right to fear a redistribution of wealth and the social revolution that would follow; just three years later, the "True Levellers," also known as Diggers, squatted on the common land at St. George's Hill and tilled soil they did not own. One of their number, Gerrard Winstanley, recorded the communist impulse behind this simple act:

[Men] buying and selling the earth . . . saying *This is mine* . . .
restrain other fellow-creatures from seeking nourishment from
their mother earth. . . . So that he that had no land was to
work for those, for small wages, that called the land theirs; and
thereby some are lifted up into the chair of tyranny and others
trod under the foot-stool of misery, as if the earth were made
for a few; not for all men.

Class hierarchy and oppression follow from "particular propriety," or,
as we refer to it, "private property."

No man can be rich, but he must be rich either by his own labours, or by the labours of other men helping him. If a man
have no help from his neighbour, he shall never gather an estate of hundreds and thousands a year. If other men help him
to work, then are those riches his neighbours' as well as his;
for they be the fruits of other men's labours as well as his own.
(Quoted in Brailsford, *Levellers*, 661)

The purpose of the debates at Putney was to produce a set of proposals to put before Parliament, proposals representing the view of
the army as to how England should be governed. Cromwell's relatively conservative views meant that he had the difficult task of making sure that the radical proposals were not affirmed by the army,
while still maintaining control over and keeping the support of that
army.

In November 1648 the army sent to Parliament a document condemning the king's policies, exposing the folly of attempting to treat
with him, and demanding that he "be speedily brought to justice, for
the treason, blood and mischief he's therein guilty of."[3] The Presby-
terian-controlled House of Commons refused to discuss the document
and the army moved to London to enforce its demands. To gain control of the government for the army, Colonel Pride's troops purged
Parliament of all but a remnant of the members, by arresting the others or forbidding them entry. These remaining men, the Rump Parliament, passed an ordinance whereby a court of commissioners was
created to try the king. (Of the 135 commissioners asked to serve,
many never appeared at the trial.) Charles refused to recognize the

court (and in this he was right; he was above the law). Nevertheless, the trial went forward; the king was sentenced to have his head severed from his body and fifty-nine of the commissioners finally set their hands to his death warrant. He was beheaded on January 30, 1648/9, in the final, wrenching moment of the Civil Wars.

By mid-March the House of Commons had abolished the monarchy and the House of Lords. Unfortunately for the new Commonwealth, its birth coincided with the third year of poor harvests, rapidly rising food costs, and very high unemployment (which its birth may have contributed to). The economic picture for landless workers, outservants, paupers, and cottagers was very bleak throughout the second half of the seventeenth century. The export of corn kept prices high, and wage laborers' loss of their small plots meant that they no longer had that cushion for times of unemployment. One news correspondent reported that "a third part of the people in most parishes" needed poor relief, though not all received it, and another reported from Somerset on the leveling opinions of the troops quartered there and added that the poor would soon become as radical as the army (Brailsford, *Levellers,* 465–466). In London, the "mob" grew, constituted in part by casual laborers who had drifted into the city in the attempt to make a living. There they could hope to escape enforcement of the Poor Law and the Act of Settlement (which provided that unwelcome newcomers be sent back to their last place of residence).

Too, the nation was shocked by the beheading of the king; many went into mourning for him. Foreign governments were horrified. The Scots declared Prince Charles to be heir to the throne and it appeared that the Protestant gentry in Ireland might unite with the confederated Catholics against their English oppressors.

John Lilburne had worried that once the "grandees" took control of England, the reforms they had promised to consider would be forgotten. He was right. The Rump had no intention of dissolving itself and calling elections for a new Parliament; it failed to act on his proposal that it abolish tithes; and the press-gang was already at work on behalf of the navy. Soon the Levellers organized a petition among the troops of the army against martial law and the use of troops as police to enforce censorship of the press. This prompted the general council of officers to request legislation that punished civilians who "breed division in the Army"; the council forbade soldiers to meet and discuss petitions, and it limited their right to petition at all. Shortly thereafter, Lilburne published a pamphlet in which he called for a strong Parliament, as opposed to the strong Council of State it had just elected, and warned against the threat of tyranny posed by an army whose commanders maintained their power continually. When Lilburne refused a well-paid post offered him by one of Cromwell's men, the Rump declared him guilty of high treason for his pamphlet, on the grounds that it was "highly seditious and destructive to the present government" and tended "to division and mutiny in the Army." The next day, not only Lilburne, but also the other Leveller leaders, Overton, Walwyn, and Prince, were arrested. By only one vote, the Council of State supported Cromwell's motion to commit them to the Tower (Brailsford, *Levellers*, 471–482).

Petitions poured in immediately. The first, on behalf of a fair trial for the four men, was signed by ten thousand Londoners; a second argued that they were really harassed for trying to "reduce the military power to a real subordination to the civil authority" (quoted in Brailsford, *Levellers*, 486). But it was the women, yet again, whose demonstrations caused the greatest notoriety. On Monday morning, April 23, 1649, several hundred women brought to the House a petition signed by ten thousand women. With an eye to the essentials— food, jobs, and taxes—the women complained in their petition, "Trading is utterly driven away, all kinds of Provision for Food at a most excessive rate, multitudes ready to starve and perish for want of work, employment, necessaries, and subsistance; Tythes, Excise, Monopolies continued to the extreme disheartning of Tillage and Trade, Taxes more and more than ever, and those rigorously executed." They charged Parliament with tyranny and demanded that the Leveller leaders be released or tried according to the law of the land. In particular, the petition argued that Parliament had to show Lilburne's pamphlet to be treasonable on grounds other than that Parliament declared it so. They entreated Parliament to

> be very wary in making any thing to be treasonable, or a capital offence, that is not essentially destructive to civil Societie: then which we know nothing more, then the exercise of an arbitrary Power, or continuance of Authoritie Civil or Military, beyond the time limited by Trust or Commission, or the per-

verting of either to unjust bloudy, or ambitious ends; things which our said Friends, with others, have much complained of, and for which principally we beleeve their lives are so violently pursued.[4]

For two days the House refused to take the petition, and on Wednesday several lively confrontations occurred, although the order of events is unclear. According to news sheets of the day, the women came to the door of the House to deliver their petition, but "*as women* were sent doun again." Finally, "The House of Commons sent out the Sergeant at Armes to the women to fetch in their Petition," but after debate, the House sent him again to tell them that

> Mr Speaker (by direction of the House) hath commanded me to tell you, That the matter you petition about, is of an higher concernment then you understand, that the House gave an answer to your Husbands; and therefore that you are desired to goe home, and looke after your owne businesse, an[d] meddle with your huswifery. (Quoted in Higgins, "Reactions," 202 and 203; emphasis mine)

According to a royalist newspaper, some of the troops guarding the House cocked their pistols and pointed them at the women while others threw squibs at them. Twenty women gained the lobby of the House only to be told by one of the members to go home and wash their dishes. Their leader replied, "we have scarce any dishes left to wash." And upon being told that it was strange for women to petition, this gentlewoman responded that "It was strange that you cut off the King's head, yet I suppose you will justify it." At some point, Cromwell himself apparently passed through their midst and promised "law" for the prisoners, only to be told that "if you take away their lives or the lives of any, contrary to law, nothing shall satisfy us, but the lives of them that do it, and Sir, we will have your life if you take away theirs" (quoted in Brailsford, *Levellers,* 487).

We can see the close connection between religion, general politics, and gender politics in the way the Leveller women went about gathering signatures for their petitions. One news sheet noted that on Sunday, April 22, 1649,

> a Petition was promoted at Severall Congregationall Meetings in and about the City of London, for women to subscribe to a

Petition to be presented to the House of Commons the next morning in behalfe of Lieut. Col. John Lilburne, and the rest of their friends in prison, in some places many signed it, in other places none at all, and in some places it was disputed. (Quoted in Higgins, "Reactions," 200)

Each ward and division of the City of London had a Leveller woman appointed to gather signatures, and the sectarian and Independent congregations to which most Leveller women belonged were good places to do it.

The execution by court martial of Robert Lockyer[5] provided the occasion for a huge march at his funeral and was of great concern to the women in their May 1649 petition to Parliament. According to Brailsford, ten thousand women signed, and a thousand delivered, a petition complaining that Lockyer had been court-martialed in peacetime and demanding justice for Lilburne and the other Leveller prisoners. In this petition, the women claimed political equality with men:

Since we are assured of our creation in the image of God, of an interest in Christ equal unto men, as also of a proportionable share in the freedoms of the commonwealth, we cannot but wonder and grieve that we should appear so despicable in your eyes as to be thought unworthy to petition or represent our grievances to this honourable House. Have we not an equal interest with the men of this Nation, in those liberties and securities contained in the *Petition of Right,* and other good Laws of the land? Are any of our lives, limbs, liberties or goods to be taken from us, no more than from men, but by due process of law and conviction of twelve sworn men of the neighbourhood? And can you imagine us to be so sottish or stupid as not to perceive or not to be sensible when daily those strong defences of our peace and welfare are broken down and trod underfoot by force and arbitrary power?

Would you have us keep at home in our houses, when men . . . as the four prisoners . . . are . . . forced from their Houses . . . to the affrighting and undoing of themselves, their wives, children, and families? Are not our husbands, our selves, our children and families by the same rule as lyable to the like unjust cruelties as they? . . .

No. . . . Let it be accounted folly, presumption, madness
or whatsoever in us, whilst we have life and breath, we will
never . . . cease to importune you. . . .

And therefore again, we entreat you to review our last petition. . . . *For we are no whit satisfied with the answer you gave
unto our Husbands and Friends,* but do equally with them remain lyable to those snares laid in your Declaration. (Quoted
in Brailsford, *Levellers,* 317, and Higgins, "Reactions," 217)

The women were all the more offended by the refusal of Parliament
to read their petition in that they had given their money, goods, and
labor to the Parliamentary cause. Cromwell's army was occasionally
referred to as the "thimble and bodkin army" because "[u]nusual voluntary collections were made both in town and country. The semp-
stress brought in her silver thimble, the chambermaid her bodkin, the
cook his silver spoon, the vintner his bowl into the common treasury
of war. . . . And observed it was that some sort of females were freest
in those contributions, as far as to part with their rings and ear-
rings."[6] Too, the women of London and elsewhere had helped to build
fortifications and dig trenches in 1642 and 1643. In the cities and
towns of England, women (on both sides) had engaged in siege work,
become garrison commanders in their besieged houses, acted as cou-
riers and spies, assisted in escapes, and occasionally become soldiers
(Higgins, "Reactions," 220–221).

These remarkable events and the extraordinary political claims of
the women, particularly their claims to "speak and act for themselves,
to have an existence and rights apart from their husbands, and not
to be included automatically in an answer that had been given to their
husbands" (Higgins, "Reactions," 218), made clear their radical gender politics as opposed to the gender ideology of most men (and probably women) at that time. Their petitions and demonstrations threatened not only private gender politics but public political arrangements
as well. Responses to them reveal that their activities were perceived
as a threat to proper decision-making authority. The words of the
Speaker and individual members of Parliament, as well as the news-
papers, proclaimed that women cannot understand state matters; that
by law husband and wife are one and that one is the husband; and
that women do not have the legal right to petition. Some news com-

mentators merely represented the gender status quo: the petitioners had been told by the Speaker of the House "in effect, that they should go home and spin; it being the usuall work of women either to spin or knit and not to meddle with State Affaires." Others, of course, could not imagine gender equality: the petitioners must be demanding, "let women weare the breeches." The only possible place for women in civil society is at home, not at Parliament: "It is fitter for you to be washing your dishes, and meddle with the wheele and distaffe"; "we shall have things brought to a fine passe, if women come to teach the Parliament how to make Lawes"; "[i]t can never be a good world, when women meddle in States matters . . . their Husbands are to blame, that they have no fitter imployment for them" (quoted in Higgins, "Reactions," 212–213).

The Republic did not last long; the Rump Parliament refused to cooperate with Cromwell and he finally dissolved it in 1653 by force of arms. Republican bitterness against Cromwell for this action never abated and returned to ruin his later Parliaments. Though the army had long argued for a more democratic election of members, army leaders now picked 140 representatives from among those nominated by the Independent churches in each county and by military officers. The Barebones Parliament (so called after Praise-God Barebon, one of its new members) was a zealous body, indeed. It voted for relief for debtors and care for lunatics; to abolish the court of chancery and to establish civil marriage. It also resolved to abolish the patronage system by which clergy were chosen, but when it became apparent that the saints might go on to abolish the state church and, perhaps, tithes (even though tithes now went to Presbyterians, Baptists, and Independents rather than to Anglicans), the army persuaded its Parliamentary friends to abdicate their powers in favor of Cromwell. He would be Lord Protector under England's first written constitution, the Instrument of Government, drawn up by officers of the army.

The threat to the state that the radicals were perceived to pose was reflected in Cromwell's opening speech to the Barebones Parliament. He exclaimed that the Levellers had sought to destroy the classes into which society was divided:

A nobleman, a gentleman, a yeoman; the distinction of these:
that is a good interest of the nation, and a great one! The natu-

ral magistracy of the nation, was it not almost trampled under foot, under despite and contempt, by men of Levelling principles? I beseech you, for the orders of men and ranks of men,
did not that Levelling principle tend to the reducing all to an
equality? Did it consciously think to do so; or did it only uncon-
sciously practise towards it for property and interest? At all
events, what was the purport of it but to give the tenant as lib-
eral a fortune as the landlord. . . . And that this thing did and
might well extend far is manifest; because it was a pleasing
voice to all Poor Men, and truly not unwelcome to all Bad
Men.[7]

He went on to attack those who "abused" liberty of conscience by
claiming that a magistrate could not punish men for holding any "notions" (as opposed to engaging in acts); those who believed in the
Fifth Monarchy (the imminent end of the world after an apocalyptic
battle against the Antichrist and then Christ's personal rule); those
who thought that anyone who wished could become a minister and
preacher; and, in short, those who wished to overturn everything—
the radical sectaries. Nevertheless, Cromwell supported religious toleration as guaranteed in the Instrument of Government. But the conservative Puritans and moderate Presbyterians of Parliament wished
to persecute those whom it deemed guilty of blasphemies and heresies.

As political arrangements shifted dramatically between 1649 and
1660 from republic to Protectorate and back to monarchy, the religious, social, political, and economic aspirations of the radicals were
increasingly disappointed. But though they were suppressed, the radicals were far from annihilated, and they remained a threat to the increasingly conservative political order. England was in fact governed
by a military dictatorship from January 1655 until the Restoration.
Cromwell died in 1658 and was succeeded by his son, Richard, who
was soon overthrown by the army. Charles II returned to England on
May 25, 1660, accompanied by one of Boyle's brothers, among others,
and two weeks later, Robert Boyle signed a document declaring his
loyalty to the king, as required by the Declaration of Breda.[8]

8

Boyle's Gender Politics

Many of the essays Boyle wrote during the late 1640s express strong opinions concerning the proper role of women; they show us that he was eager to reinforce a femininity more consonant with bourgeois ideals than with traditional, aristocratic ones or with upstart, sectarian ones. Thus, many of his letters excoriating Corisca strike at the courtly image of femininity as artful (altering the natural), "bold," "immodest," and bordering on unchaste (an image which was not necessarily an accurate depiction of aristocratic women in any case). On the other hand, he was very concerned to undercut an emerging counter-image of femininity by depicting it, too, as "bold," "immodest," and bordering on unchaste. This was the image of sectarian women, boldly speaking out in public, affronting their own "proper modesty" by publicly preaching, prophesying, and praying, traveling abroad to preach the gospel without male protectors, and leaving husbands who refused to join the company of "the saints."

One of his earliest epistolary essays attacks the upper-class and court practice of sending babies away to wet-nurses. "The Duty of a Mother's Being a Nurse Defended," dated August 15, 1647, takes as its thesis that babies should be breastfed and brought up by their own mothers. Typically, these epistolary essays state a thesis and adduce arguments from piety, conscience, and science to persuade the reader. In this essay as in others, Boyle sets up a fictional character who exemplifies the vices he is concerned with and against whom he argues; thus, "The Duty of a Mother's Being a Nurse Defended" includes a lengthy tirade against "Corisca" for "being a nurse by proxy." When she sends her child out to a wet-nurse, she omits her duty, he charges, for the law of nature and of God allots nursing to mothers—who, Boyle says, do not deserve the name "if their sensuality (and its regretted effects) have given them all the title they have to it." Breastfeeding is the purpose of "those snowy hills," even though we find no positive command in scripture (because it would have been superfluous). Boyle also adduces several medical arguments: "physicians

unanimously pronounce that . . . the mother's milk is the most pro-
portionate to the infant's temperament and constitution (being made
of the same blood of which the child was fashioned and furnished in
the womb) and as most natural to it, most healthful for it." Moreover,
sending it out (throughout this essay Boyle refers to the infant as "it")
exposes it to diseases that are common in mercenary nurses because
they have what Boyle calls a "mettled impatience." And in any case,
the baby will acquire the nurse's vices and her passions by drinking
her milk. Too, children are not as well tended by nurses as by their
mothers, since the nurse is motivated by wages and not by affection. Thus, they get worse education and "breeding," for they learn
"meanesses and unhandsome words and tricks."

Perhaps motivated by his own experience as a child when he was
sent out to a "country nurse,"[1] Boyle takes the first person in his closing words: I have no obligation to a mother who is not motherly, he
says, who

> designelessly conceives, gives birth necessarily, to be rid of a
> pain, sends me abroad when I most need her tenderest assidui-
> ties and kindness, uses me . . . for her sport. . . . parts with me
> as a stranger and loves me but under the notion of an ape.[2]

Real mothers keep their children with them, breastfeed them, take
them seriously, and truly love them.

Wearing make-up, black patches, and low-cut dresses was associated with upper-class and court circles, so Boyle's attack on these
practices was addressed to upper-class women. But Dolan tells us that
popular recipe books included recipes for cosmetics along with those
for dishes and medicines, indicating the interest of middle- and lower-
class women in cosmetics. Boyle's letters discoursing against make-up,
patches, and low-cut dresses belong to the debate over cosmetics extending from the Renaissance through at least the late eighteenth
century. Boyle's attacks were, like many others, concerned for wom-
en's health and for their souls, especially since the use of cosmetics
was deemed to usurp God's prerogative as creator. But most attacks
were conjoined "with anguished concern over manliness." Dolan
points out that in displacing their anxiety onto women's wearing
make-up, authors were silent about men's wearing makeup and hid
their concern that doing so made them effeminate.[3] This issue be-

comes sharper to late-twentieth-century readers when we look at portraits such as Boyle's, depicting a figure wearing make-up and a wig with long, carolingian curls.

Boyle's objections to these vices are set out in an epistle addressed to Mrs. Dury (probably the wife of his friend John Dury, advocate of Protestant unity and Baconian science).[4] Opposing first the use of make-up, Boyle begins:

> I think I may with as little rudeness as much justice take this liberty to vent my choler against the vanity of Corisca, which I am unable (any longer) to endure as I find unpersuadable to desist for (in the first place) I think it will be no news to tell you that she Paints (or rather plasters).

She should be dissuaded by the fact that Jezebel, "whose crimes have made her the hyperbole of wicked women," is the only instance of "painting" mentioned in the scriptures, but Corisca does not really believe in the authority of scripture. Therefore, Boyle says, he must appeal to her conscience; isn't she ashamed "to do what she would blush to have known," and why does she not "detest the act of which she need fear the very discovery"? Moreover, "all the vermilion they daube upon their cheeks turns to Inke upon their reputations: for few men will be found refractory to believe that . . . these ladies desire this complexion to please others than their husbands: and that they are ambitious of Beauty for no very noble ends: that blush not to seek for it by such unhandsome means." And as they grow old, they will either have to continue painting, in which case their cheeks will seem "some 30 years younger than their hands and breasts," or they will stop painting and expose themselves to derision.

His appeal to considerations of health turns on the dangers of mercury, which

> perverted to such an use makes the abuse its own rod, making amends for the sweetness it gives the face by unperfuming of these and giving no snowiness to the cheeks which it takes not away from the teeth.

And his appeal to piety argues from "that which a Puritan would least forgive in Corisca's painting . . . which is that it favors, partly, of repining at Providence as if it had given her a face unfit for the ends (best known to itself) of her creation: and partly of pride as if God had given her a face worse than she preferred and therefore she would mend the work of her Maker."

Boyle also excoriates black patches or "shreds of black taffety" worn on the face:

> painting and black patches are like hypocrasy and profaneness: both antagonists to piety and (by consequence) both sins; but differing in this that the one has the bashfulness to conceal that impiety which the other has the brazenness to proclaim. For so both she that Peters [i.e., paints] and she that patches are both guilty of painting but they are unconformant in this; that the one endeavors to disguize what the other has the impudence to own. Tis such [crossed out: females] semi-Adamites as Corisca that have so cheapened the [crossed out: price] virtue of your sex/ that whereas in the world's infancy men were wont to give portions to their wives, now days they require/expect portions with um and oftentimes marry the woman but as a provise annexed to the possession of the dowry. (BP, xxxvii, 204)

Finally, against the wearing of low-cut dresses, Boyle has this to say:

> But for the present I shall only entreat her to consider in the first place: what wanton flames this snow of her breasts kindles in those of others (which is to make the livery of innocence it-self the bellows of a crime) whose concupiscences though she be not guilty of satisfying, yet she is of provoking/raising. And to say she does it [un]designedly is as good an excuse as the reason that promiscuously casting down tiles from the top of the house into the street with a purpose only to rid himself of the tiles, not of the men: who if he chance to mischief the by-passers is not excused by his design but punished for his indif-ferention. Since the harmlessness of the intention can never authorize unwarrantable means. In the next place would be considered the murder this behavior commits upon her own reputation: her breasts being not more naked than men's

tongues will be active in censuring them. . . . the last [objection] I have to mention which is that Corisca, by this lure of a bare neck . . . exposes her chastity to unnecessary dangers, by usurping the devil's office, and tempting others to tempt her. Neither will it satisfy to object that by the same reason they should be obliged to mask their faces; as parts much more tempting than their breasts; for it may easily be replied/ answered that (besides that the handsomest parts are not always the most seducing) the face was given them to be known by and so ought not to be veiled: but the breasts were never designed for any such note of discrimination but ordained to give such with which most of them I'm sure will never do. And though some will not be ashamed to give, the truest motive (tho seldomest alleged pretence) which is that this discovery makes them appear more lovely: to this it may be answered; first by demanding what fires they think this loveliness can kindle, and then by replying that by the same reason it were equally lawful for them to bare some other parts of their body, which though more conducing to that end, nature herself teaches the most barbarous people to conceal.

Throughout this letter, Boyle's fundamental concern is that women be chaste and modest, not bold and impudent. Wearing make-up, patches, and low-cut dresses were, thus, ways that women could fail of virtue.

Sectarian women also failed to fulfill the bourgeois virtues; in the name of religion, these "saints" were not silent, but spoke out in public. And they were publicly disobedient. We find Boyle's image of the proper woman, the bourgeois saint, in Theodora, the heroine of his story *The Martyrdom of Theodora and Didymus,* written in 1648–49. His purpose, he says, is to make the story of the martyrdom of Theodora and Didymus into a romance depicting their love for one another. Theodora, a beautiful Christian virgin, refused to sacrifice to the pagan gods of the Roman emperors and was condemned by the president of Antioch to life in a brothel. When her admirer, Didymus, hears this news, he rushes to the brothel and persuades Theodora to don his soldier's uniform and escape. They exchange clothes, and she departs, leaving Didymus in her place "in drag." He is found and is sentenced to death, but upon hearing of his sentence, Theodora rushes off to the tribunal to plead that she be executed in his stead. The president sentences them both to death and they are executed, but not before they engage in several disputes, including one over who should have the honor to die for whom and a theological one over whether their bodies, when resurrected, will have age, sex, and so forth (Didymus hopes they will) or whether these qualities will be transcended (Theodora expects that they will be).

Jacob reads the story as a response to the "enormous radical, sectarian activity" during the period between 1648 and 1655 and describes it as "a thinly disguised allegory about the threats to true religion presented by the alliance between the New Model and the sects at the moment of its [the story's] composition."[5] Boyle took the view that the army and the sectaries had destroyed the civil government, and his story recommends the doctrine of passive obedience to those who now found God's laws in conflict with man's. Jacob finds a precise statement of Boyle's own position in Didymus's words: "if we cannot yield an active obedience to the commands of the civil sovereign, we do not refuse him the utmost we can consent to, which is passive obedience: and when our consciences permit us not to do those, to us, unlawful things, that he commands, they enjoin us to suffer unresistedly, what ever penalties he pleases to impose."[6] Theodora and Didymus exhibit this principle in their life and their death.

But Theodora is more than a model Christian saint; she is a model of saintly womanhood: chaste, silent, and obedient. The brothel is literally for her a fate worse than death; in fact, her modesty and chastity make the mere thought of it horrible, but her obedience makes her readily choose it over death:

the infamy of this detestable place was that, which I could not think on, without the utmost horror and indignation; and not only my sex and breeding, but even the dictates of more than one virtue (modesty and chastity) concurred to heighten my abhorrence of it: but on the other side, I remembered, that I did not vow obedience to God with any exceptions or reserves. . . . And I thought, that, perhaps, Providence had led me into this distress, to give me an opportunity of shewing, that I could do more than die for Christ.

Her modesty is a sign not only that she is chaste, but also that she follows the injunction to silence. Boyle displays this aspect of her modesty again and again in the way that he prefaces her speeches. She never willingly speaks in public and does so only under the greatest duress; she would (almost) rather die. When she saw her brave lover at the tribunal, Boyle tells us,

> This moving sight so affected the generous and compassion-
> ate Theodora, that though in so publick place and manner
> she could with less reluctancy die for Didymus, than she
> could plead for him; yet her gratitude surmounting her
> bashfulness, after some conflict within herself, she made
> towards the tribunal.

And having given her speech, Boyle reinforces her preference for silence: "Here Theodora paused a while, partly to recover from the disorder so unusual and difficult an effort of her modesty had put her into."

Her final speech is prefaced with the same emphasis on her modesty. She is taken to the place of her execution and, after looking around,

> she paused a while, to recover from some disorder, that she
> was put into; not so much to see herself invironed with guards,
> as surrounded with gazers. And then, though her bashfulness
> made it more uneasy to her to speak to the assistants, because
> her speech must be a public one, than because it must be her
> last; yet with a voice and gestures, wherein the modesty of a
> virgin, and the courage of a martyr, were happily tempered,
> she addressed herself to those that were about her. (*W,* V, 267,
> 287, 309)

As Jacob says, Theodora offers a counter-example to those holding radical sectarian religious and political ideals, and shows us Boyle's concern about the sectarian threat to civil government and Christianity. But Boyle was also concerned about the activities of sectarian

women, and in Theodora he offers a woman whose religious and political ideals do not include public speech and public activities (Boyle
makes clear again and again her reluctance to appear and to speak
publicly). This is a saint whose femininity stands in contrast to the
sectarian woman "saint," who is not chaste, silent, modest, or obedient to public or private authority.

Theodora also expresses Boyle's belief that one who would serve
God with "undistracted devotion" must remain chaste in the traditional sense—unmarried and celibate (as opposed to the emerging
bourgeois sense of chastity as fidelity within marriage). And as we
saw in chapter 1, he further believed that masculine devotion to science requires the strict chastity of women.

9

Boyle's Background Reading

This chapter introduces the different intellectual traditions Boyle studied, including work in the natural-magic tradition, the new mechanical philosophy, and the Aristotelian tradition. Thus, we find that Boyle was well read in the hylozoic tradition and aware of its association with radical as well as reformist social and political movements.

Robert Boyle's surviving correspondence reveals the political, economic, and physical threats facing him during the Civil War as he undertook the scientific studies for which we remember him. Having arrived at Stalbridge in March 1646, he wrote to his sister, Lady Katherine Ranelagh,

> My stay here, God willing, shall not be long, this country being generally infected with three epidemical diseases (besides that old leiger sickness, the troop-flux) namely the plague, which now begins to revive again at Bristol and Yeovil six miles off, fits of the committee, and consumption of the purse.
> (*W*, I, xxx)

Here Boyle makes light of very serious concerns; dysentery was a constant problem and the bubonic plague later ravaged the country (and London in particular until the great fire of 1666). "Fits of the committee" probably refers to the Parliamentary committees set up to gather information on royalists as well as to sequester their estates. The Boyle family was not immune from these committees; Robert's brother, Richard, was a royalist and only a few months later (when the royalist forces yielded at Oxford) Boyle wrote to tell Richard that he could "compound" for his estate. Richard did so, paying a fine to Parliament set at one-tenth of its value.[1] And Robert Boyle himself was imprisoned not long after, for in October he wrote to his old tutor,

> I were guilty of an ingratitude great as the favour I have received, if I did not acknowledge a great deal of mercy in God's dispensation towards me; which truly hath been so kind, as

oftentimes to work my good out of those things I most feared
the consequences of, and changed those very dangers, which
were the object of my apprehension, into the motive of my joy.
I was once a prisoner here upon some groundless suspicions,
but quickly got off with advantage.

His Parliamentary connections no doubt provided the advantage, but
Boyle went on to explain that he felt himself caught between a political rock and a hard place:

I have been forced to observe a very great caution, and exact
evenness in my carriage, since I saw you last, it being absolutely necessary for the preservation of a person, whom the unfortunate situation of his fortune made obnoxious to the inju-
ries of both parties, and the protection of neither.

And the "consumption of the purse" was explained by his having

been forced to live at a very high rate, (considering the incon-
siderableness of my income) and, to furnish out these ex-
pences, part with a good share of my land, partly to live here
like a gentleman, and partly to perform all that I thought expedient in order to my Irish estate, out of which I never yet received the worth of a farthing. (*W,* I, xxxiii–xxxiv)

Despite these worries, his travels, and visits to and from his neigh-
bors, Boyle found the time to write many of his non-scientific works,
including some of his *Occasional Reflections,* short essays drawing a
moral from everyday events; romances (such as those his brother
Broghill produced so successfully), most notably *Theodora;* parts of a
projected treatise on ethics titled "Aretology or Ethical Elements"; letters; religious essays; and a theological treatise in romantic form, referred to as *Seraphick Love.* And he was beginning to devote himself
to experimental work in chemistry and medicine. A letter of 1646/7
to Lady Katherine Ranelagh indicates his interests:

That great earthen furnace whose conveying hither has taken
up so much of my care, and concerning which I made bold
very lately to trouble you, since I last did so, has been brought
to my hands crumbled into as many pieces, as we into sects;
and all the fine experiments, and castles in the air, I had built

upon its safe arrival, have felt the fate of their foundation.
Well, I see I am not designed to the finding out of the philosophers stone, I have been so unlucky in my first attempts in
chemistry. My limbecks, recipients, and other glasses have escaped indeed the misfortune of their incendiary, but are now,
through the miscarriage of that grand implement of *Vulcan,* as
useless to me, as good parts to salvation without the fire of
zeal. (*W,* I, xxxvi)

But by 1648 he had performed "many of the classical chemical experiments on metals" and in the summer of 1649 he mentions "my
new-erected Furnaces."[2] These (along with his illnesses), he tells
Lady Ranelagh, made him oblivious to all else:

for Vulcan has so transported and bewitched me, that as the
delights I taste in it, make me fancy my laboratory a kind of
Elysium, so, as if the threshold of it possessed the quality the
poets ascribed to that Lethe their fictions made men taste of,
before their entrance into those seats of bliss, I there forget my
standish and my books, and almost all things. (*W,* VI, 49–50)

His early writings as well as his interest in chemistry reveal the influence of Hartlib's Invisible College, with whom he kept in close touch
through letters and constant visits to London. That influence also included the thinkers whose works Boyle read at the time:

by the help of anatomical knives, and the light of chemical furnaces, I study the book of nature, and consult the glosses of
Aristotle, Epicurus, Paracelsus, Harvey, Helmont, and other
learned expositors of that instructive volume.

And he came to doubt the solidity of Aristotle's principles by the opposition of "the Chymists in generall and great store of Moderne
Physitians but by acute and fam'd Philosophers and the sect of the
new Copernicans, Telesius, Campanella (and his Ingenious Epitomist
Comenius), Bacon, Gassendus, Descartes and his sect" (*W,* I, 262; and
BP, XXXVIII, quoted in Boas, *Robert Boyle,* 27). Boyle was reading,
then, in at least three traditions of natural philosophy: Aristotelian
and neo-Aristotelian hylozoism (Aristotle and Harvey), natural-magic
hylozoism (Paracelsus, Helmont, Telesio, Campanella, and Comenius)

and the new mechanical philosophy (Epicurus, Descartes, and Gassendi).

Paracelsus, Helmont, Telesio, Campanella, and Comenius belong to the natural-magic tradition, whose sources included works attributed to Hermes Trismegistus, translated from Arabic into Latin by Ficino in 1463 at the behest of Cosimo de' Medici. Although these works were probably written between A.D.100 and 300, Renaissance thinkers believed that Hermes had been an Egyptian priest at the time of Moses. These writings concerned astrology and the occult sciences, the secret virtues of plants and stones, and sympathetic magic, including the making and use of talismans for drawing down and storing the power of the stars. Through an ascetic, disciplined way of life, the mage contemplates the cosmos as it is reflected in his own mind and reaches ecstatic illumination as his soul ascends through the planetary spheres and is joined to the divine.[3]

The Hermetic writings do not offer a systematic, consistent set of doctrines, but the central location of man in the cosmos is clear throughout. Man was made in the image of God and given dominion over all creatures; but he was also divine, with divine creative power. However, he leaned down through the armature of the spheres, tore open their envelopes, and showed himself to Nature; she fell in love with him and he took on a mortal body, thereby becoming subject to the influence of the stars. He nevertheless kept his divine, creative, immortal essence—which he recovers, along with his knowledge of and power over creation, when he achieves illumination (Yates, *Giordano Bruno*, 28). In his "Oration on the Dignity of Man," Pico della Mirandola describes the work of the mage who,

> in calling forth into the light as if from their hiding-places the
> powers scattered and sown in the world by the loving-kindness
> of God, does not so much work wonders as diligently serve a
> wonder-working nature . . . and having clearly perceived the re-
> ciprocal affinity of natures, and applying to each single thing
> the suitable and peculiar inducements . . . brings forth into the
> open the miracles concealed in the recesses of the world, in the
> depths of nature, and in the storehouses and mysteries of God,
> just as if she herself [Nature] were their maker.[4]

Man can bring forth miracles just as Nature herself does. The early modern mage was to recover lost knowledge by studying ancient myths, the *Corpus Hermeticum* and the Cabbala.

Animism was central to the Hermetic world-view. Yates tells us that according to the *Corpus Hermeticum*,

> all this great body of the world is a soul, full of intellect and of God, who fills it within and without and vivifies the All.

And again,

> All that is in the world, without exception, is in movement, and that which is in movement is also in life. Contemplate then the beautiful arrangement of the world and see that it is alive, and that all matter is full of life.

And finally, in a passage suggestive of pantheism, Tat asks Hermes, "Is God then in matter, O Father?" To which Hermes replies,

> Where could matter be placed if it existed apart from God? Would it not be a confused mass, unless it were put to work? And if it is put to work by whom is that done? The energies which operate in it are parts of God. Whether you speak of matter or bodies or substance, know that these things are energies of God. . . . In the All there is nothing which is not God. (Quoted in Yates, *Giordano Bruno,* 31, 34)

These basic ideas, and many others, were taken up in the northern European Renaissance, particularly in the form given them by the Florentine priests Marsilio Ficino and Pico della Mirandola. These thinkers attempted to Christianize the Hermetic writings and systematized many of the ideas set out in them. For Ficino (in the tradition of Platonists, Stoics, and Neo-Platonists) the world is one animal; like man, it has a soul and a body as well as a third thing, a spirit which is the medium between the soul and body. In man, this spirit appears to be a corporeal vapor made of the four elements, but the spirit of the world is made of a fifth element or quintessence (Aristotle's ether). Ficino says this *anima mundi* "is a very subtle body; as it were not body and almost soul. Or again, as it were not soul and almost body. . . . It vivifies everything everywhere and is the imme-

diate cause of all generation and motion."[5] Generation and preservation of bodies as well as sense perception in man and animals are all
possible because the *anima mundi* makes animal spirits more akin to
the soul. And it is enough like our spirit that we can absorb it and it
will nourish our spirit; this can be done by ingesting such things as
wine, white sugar, and gold, or by smelling cinnamon or roses—all
things containing the soul of the world in a fairly pure form (Walker,
Magic, 55, 13).

It is also possible to attract the spiritual influence of the stars
(which include the seven planets, the sun, and the moon, as well as
the twelve signs of the zodiac) by using people, animals, plants, metals, and minerals, including gemstones, which are subject to a particular star and constitute a hierarchical series with it. (Later, these
were referred to as "celestial magnets.") This is possible because of the
sympathies among the things in each series. Sympathies are the har-
monic proportions among the things; each planetary sphere makes a
musical sound and anyone born under the influence of a planet has
the same numerical proportions. Animals, plants, or minerals subject
to the planet also have numerical proportions similar to it; therefore,
by taking proper measures (wearing certain planetary colors, strew-
ing the right herbs around, singing the right songs, etc.), it is possible
to provoke the spirit of a planet in the same way that "a vibrating
string will make another, tuned to the same or a consonant note, vi-
brate in sympathy" (Walker, *Magic,* 14). Sympathy and antipathy
were fundamental physical principles used to explain all actions, such
as magnetic attraction and repulsion and, most importantly, the medicinal actions of animals, herbs, and minerals, music and talismans.

Ficino was a physician as well as a priest and held the normal assumptions about the body, health, and disease, e.g., "that the signs
ruled different parts of the body, that different bodily temperaments
were related to different planets" (Yates, *Giordano Bruno,* 62). In his
medical treatise, *Libri de Vita,* therefore, he includes uses of natural
magic set out in the *Corpus Hermeticum,* particularly the use of talismans, to cure physical and mental illnesses.

We find the same desire to use natural magic to bring people health
and long life in Paracelsus (1494–1541). For Paracelsus as for Ficino,
matter is alive; it is spirit, but spirit is understood to be very fine
matter—in fact, matter is produced through the continual solidifica-

tion of spirit. Against the Aristotelians, Paracelsus argued that individual objects and species differ essentially not because they have different chemical components, but because they have different "immanent, specific, soul-like forces." These forces are the semina and are contained in every object in nature. Here Paracelsus generalizes the hylozoic, alchemical view that metals germinate from seeds planted deep in the earth. Thus, the semina determine how all objects develop their qualities and characteristic features.[6]

Inherent in the semina are the three elements or principles, salt, sulfur, and mercury, understood not as the simplest elements which are mixed to form bodies, but as dynamic, functional guiding forces. Salt is the principle that directs material toward the solid state and gives it a certain color; sulfur, the principle that makes it more or less combustible and gives it structure; and mercury, that makes it fluid or vaporous and gives it certain virtues and powers. Although Paracelsus says that objects contain all three principles, he also says that each object is the "fruit" of only one element—understood as a spirit, alive in each thing, not as something which appears to our senses. Nevertheless, to prove that all objects contain the three principles as chemical elements, Paracelsus argued that one should burn them. Burning removes "their coarse visible covering" and exposes their "invisible kernel," the combustible, the vaporous, and the solid. "Hence the true naturalist is called 'Philosopher through Fire'. If wood is burnt, it will be resolved into its three true components: the flame—its 'sulphur', the smoke—its 'mercury' and the ash—its 'salt'" (Pagel, *Paracelsus,* 87, 88, 93, 101).

The three principles together form prime matter; for Paracelsus, prime matter is uncreated, immortal, and the original spiritual mysterious stuff which, by separation, i.e., individuation into parts, becomes the three elements. Pagel tells us that this concept of prime matter reveals Paracelsus's pantheistic tendency. The concept ultimately derived from the Stoics, who held that spirit and matter are united to form living matter, but Ibn Gebirol (1020–1070) understood this matter to be formless, i.e., undifferentiated and infinite, having no quantity or corporeality, since these already imply specification and limitation. And Gebirol's prime matter was "a fundamental source for medieval pantheism"; it came to be thought of as God, the "indivisible substance that originates, sustains and exists in all things" (*Paracelsus*, 91–92, 229, 230). This concept influenced the heretic movements of Paracelsus's time: Waldenses, Beguards and Beguines, Brothers of the Free Spirit, and many others. Paracelsus's life and work show many contacts with the ideas of these heretics and social revolutionaries:

> Certain traits exhibited in Paracelsus' life and behaviour as an individual are reminiscent of those displayed by the groups of heretics which emerged in the "autumn of the Middle Ages," notably the Beghards and the Beguines. Driven by feverish unrest, these preachers in the name of the unique "spirit of the world" found their main satisfaction in acts of agitation and in wandering from place to place through Switzerland, Strasburg and Colmar—familiar haunts in the life of Paracelsus. . . . Homeless, they made their home everywhere. Practical people, most of whom had learned a craft, they felt keenly the need for social reform and for liberation from feudal slavery. Dedicated to the other world, they nevertheless adopted the attitude of temporal rulers and did not hesitate to use threats in launching their demands for the communal distribution of goods. They preached wisdom from spiritual experience as against books, even including the Scriptures. For them there was no God outside the world. The soul of man was divine, and man's duty was to unite with it completely so that thus God might be caused to descend into man and man to ascend to divinity. (*Paracelsus*, 231n)

They also rejected the divinity of Christ and the necessity for a mediator between God and man. Like the social reformers and revolutionaries of his time, Paracelsus was critical of both ecclesiastical and secular powers, but he recognized state and church authority, within limits, and assumed that the emperor should communalize the land and the means of production. Pagel tells us that Paracelsus's "life and work was a permanent war against the privileged and mighty." He charged the rich large fees for his medical services but cured the poor for free. Thus, he sympathized with the peasants of Salzburg in their 1525 revolt against their feudal lords and barely escaped death during the struggle (*Paracelsus*, 43, 40).

His tendency to dissolve hierarchies is also found in his view of correspondences. In the traditional view, as we have seen, each planet crowns a hierarchy of people, animals, plants, minerals, and elements, but Paracelsus thought all members of the hierarchy are bound together by sympathies, so that any action brought about in the virtue or spirit of one affects all the others, including the planet (*Paracelsus*, 37, 38). Sympathetic actions are possible because all things are connected through the world spirit. Thus, the *anima mundi* makes the work of the mage/physician/priest possible: by sympathetic magic using the proper plants and minerals—or chemicals made from them—he manipulates the virtues or powers in the system to bring about human health and happiness.

Paracelsus thus attempted to make himself the ideal priest-physician exemplified in the work of Ficino. Just as the mage uses magic to capture the heavenly gifts hidden in natural objects for the benefit of human happiness and health, so the priest-physician receives his power to heal from celestial inspiration, so that he knows how to compose medicines under the most propitious stars using the correct ingredients. This natural magic is possible because man unites in himself two worlds—the visible world of matter and the invisible, celestial world of action and power. He is a microcosm of the greater world, the macrocosm, for he has flesh and blood from the elements, and a celestial or "astral body" (i.e., a spiritual body—here again understood to be very fine matter, not something totally immaterial). As the microcosm of the macrocosm, man contains in himself all the constituents of the world, minerals, plants, animals, and celestial bodies; thus, he can acquire knowledge directly, not by rational consideration of objects outside the mind, but by an act of sympathetic attraction between the inner representative of a particular object in his own constitution and its external counterpart. He unites with the object and knows it intimately. For example, an herb knows how to purge, so to learn about it, one listens to it and experiences it (Pagel, *Paracelsus*, 222–223, 50, 60). Its knowledge becomes one's own.

His emphasis on illumination and inspiration was in direct opposition to Galenic medicine, practiced by followers of Aristotle. They attributed disease to too much or too little of the four humors (blood, phlegm, black bile, and yellow bile) and the four qualities (dryness, dampness, heat, and cold). The only disease, then, is distemper, though symptoms differ according to the individual's mixture of humors and the four qualities, and the job of the physician is to calculate the distemper and add what is lacking or take away what is in excess (usually by bleeding or purging). The Galenic physician uses reason; Paracelsians receive illumination.

Paracelsus opposed the humoral theory of disease, arguing that it could not explain the large variety of diseases, and his great contribution to the history of medicine was his belief that each disease is a specific entity which enters the body from outside (Pagel, *Paracelsus*, 129–130). Because man is a microcosm of the cosmos, his body is a replica of the heavens; thus, organs are in the body what the stars are in the world: they dominate life, functioning as individuals yet interacting. Disease is basically a disturbance of the interaction of the organs. Each disease is caused by a specific "seed," which can be a poison, an astral virtue, a command of God, or the product of an abnormal imagination. The disease agent acts on an equivalent substance inside the body, producing a disease complex with its specific characteristics. To remove the specific agent or disease complex, Paracelsus thought that we need specific remedies (sympathetic or homeopathic to the disease), and the search for specific remedies led him to develop chemical medicines, e.g., mercury in the proper dose and form was his specific remedy for dropsy (Pagel, *Paracelsus*, 140–142).

In some texts, Paracelsus looks for the dawning of a new "Joachimite age of the Holy Spirit" in which all secrets would be revealed and the arts and sciences perfected. And the knowledge gained, particularly medical knowledge, would be used for Christian social ends: to relieve human suffering. These views made his doctrines especially attractive to German millennialists, including Valentin Andreae and Jan A. Comenius (Paracelsus's "ingenious epitomist"). Because they thought the millennium was near, these thinkers sought "social, religious and educational reform . . . marked by the recovery of the knowledge of creatures that Adam had possessed in his innocence, and of the Adamic language which had given him power over all things" (Rattansi, "Social Interpretation," 12).

Others also took the Hermetic natural philosophy to be part of the social and religious reformation of the whole world; for example, in the writings of Giordano Bruno this view led to the heresy that Christianity was the corruption of a more ancient, true 'religion of the world,' viz., Hermeticism. We also find Hermetic natural magic in Thomasio Campanella's *City of the Sun*, describing a utopian republic led by a Hermetic magician and characterized by its encouragement of scientific investigation to improve the life of the community as well as by eugenics and by the communal ownership of property. Campanella was an advocate of Copernicus's new theory of the solar system, but he understood it in the context of a dense astrological magic derived from Ficino (see above). His utopian city is laid out in seven circles reflecting the seven planets and is "a complete reflection of the world as governed by the laws of natural magic in dependence on the stars." Moreover, the eugenics practiced in the utopia is "concerned with choosing the right astrological moment for conception, and with mating males and females in accordance with their astrological temperaments." Campanella's "Hermetic utopian vision" also influenced Jan Comenius and Valentine Andreae, whose utopian *Christianopolis* is indebted to it (Yates, *Giordano Bruno,* 369).

In the work of Helmont, natural-magic beliefs are systematized and made more coherent with his careful observations, experiments, and measurements. For Helmont, water is the one basic element, in itself inert, "empty matter" (i.e., a version of prime matter). This is not the liquid we drink, but the original water over which God "hovered" at creation. It is transmuted into different bodies with specific properties by semina or "the seeds of things." These seeds are the central agents for the generation of everything, animal, vegetable, and mineral, and for all changes in them. Pagel tells us that for Helmont, all generation begins with an image or idea. Semina are endowed with these images, or "spiritual type-plans," which they transmute into bodies. More specifically, the archeus or vital spirit within each seed imagines the idea and directs and enlivens the semen. The archeus has "knowledge of what has to be done" and directs the seed so that it transmutes empty matter into a body with specific properties.[7]

All archei are created by God, and although each is his trustee and executive, they form a hierarchy. The World Soul or "common Intelligencer" presides over all; thus, the universe itself and each thing in it is animated. Helmont distinguished among minerals, plants, animals, and men not by whether they live but by the ways they live and by their mode of perception. Minerals are not without "sense," for "innate sensation (*naturalis perceptio*)" is common to all things. "Thus,

Van Helmont speaks of a 'deaf perception' which appertains to any object . . . and also of a rudimentary *sensus* in any inanimate body. The next higher stage is vegetation and nutrition, to be followed by *sensus* and irritability, intrinsic properties of animal tissue [according to Pagel, "tissue intellect . . . enables a tissue or organ to judge or to 'know' what is good or harmful for itself and the economy of the organism as a whole"], and finally by *intellectus*—the divine *mens* [mind] which is the prerogative of man" (Pagel, *Van Helmont*, 118–119).

Each individual unit, whether a man, an organ, or a stone, is constituted of matter and spirit or semina—united, these constitute a living monad. And the archei within monads are responsible for the actions and interactions among monads, interactions occurring by sympathy and antipathy, imagination and spiritual (magnetic) attraction. Where Helmont differs from other Paracelsians is in his belief that the archeus of an object can be demonstrated *in vitro*, as an "object-specific smoke" or gas. In the concept of a gas, Helmont thought that he had found the physico-spiritual sort of substance serving as a third thing, the medium between pure spirit and inert matter. He is credited with discovering gases, and he identified fifteen, some of which, e.g., carbon dioxide, we still recognize. But he conceived of gas as original inert water which has been "signed" by a semen. It is volatile water; not water vapor, but a kind of smoke. "*Gas* becomes manifest when a solid body is made to relinquish its 'vestments,' the husk or shell that conceals its essential (spiritual) centre. This is accomplished by burning the vestments away, [by the chemist's fire], or by nature, as for example in grapes that start fermenting" when their skins are removed (Pagel, *Van Helmont*, 43, 96, 62).

Gases are not to be confused with air, which Helmont thought of as an element like water. Air can be compressed, and has the power of a fire-arm if it is put under high pressure and suddenly released. Moreover, "Van Helmont reckoned that *air could be reduced to one half of its original volume under pressure.*" He concluded from this that the other half must have consisted of space containing no bodies (Pagel, *Van Helmont*, 92, emphasis mine). Experimentally, we see that when the flame of a candle burning on top of a water surface and enclosed in a cylinder has consumed the air, the water rises in the cylinder and extinguishes the candle. Helmont argued that the otherwise empty space within the air contains magnale or spirit of the air. This magnale is also a third thing between spirit and matter, and it, not air, keeps us alive when we breathe. It is the World Soul. And because Helmont identifies the World Soul with magnale, he believes that we can measure the effects of the World Soul, e.g., we can measure how much of it is consumed by a burning candle if we measure how high the water rises in the experiment just mentioned.

Finally, the World Soul makes possible sympathetic and antipathetic effects at a distance. Sympathy, the yearning of like to join like, explains not only ordinary effects such as sympathetic vibration between strings proportionally attuned, but also "paradoxical" effects such as the attraction of iron to the lodestone, the attraction of disease entities for the perverse imagination, and the response of precious stones to sicknesses. To Helmont, all such effects are legitimate and natural, i.e., natural magic, "nature being the magician by virtue of universal sense and sympathy" (Pagel, *Van Helmont*, 11).

10

Boyle's Hermeticism, Magic, and Active Principles

Investigation reveals that Boyle had a strong interest in and some anxiety about magic and that he was not as thorough-going a mechanist as, say, Descartes.

Peter Rattansi argues that Boyle was typical of many English virtuosi of the time in his eclectic approach to the study of natural philosophy. For example, William Petty, a member of the Hartlib circle for a time and later a member of the Royal Society, also listed Descartes, Paracelsus, Campanella, and Helmont as important modern thinkers whose writings he searched. Petty was clearly influenced by the animism of the latter three thinkers, for, in a discourse to the Royal Society published in 1674, he argued for an atomic theory, but one in which atoms are understood to be male and female—a substantial distinction, he argued, and not a metaphorical one—because the Bible mentions that God created male and female, though not that he created atoms. Just as Helmont's animism was compatible with his use of measurement and experiment, Petty's animism was compatible with the mathematical nature of science, for in the same discourse he argued for "giving a mathematical character to various corpuscular explanations."[1] For Petty as for Boyle, Paracelsus, Campanella, and Helmont offered possible alternatives, along with Galileo, Descartes, and Gassendi, to an increasingly discredited Aristotelianism.

In this context, Boyle's enthusiasm for the principles of the Hartlib circle, valuing "no knowledge but as it hath a tendency to use," particularly medical knowledge, and his enthusiasm for Hartlib's plan to publish Campanella's *City of the Sun* and Andreae's *Christianopolis* (*W*, I, xxxviii) take on new significance for us. His interest in chemistry, "the science *par excellence* of the Hermetics," arose in part from "his hypochondriac self-medication," but in part from his interest in Hermeticism. We shall see that Boyle criticized Paracelsian theories in *The Sceptical Chemist*, and similarly criticized other views arising from the natural-magic tradition in later writings, but as Rattansi notes, "many of the problems that obsessively recur in his work are of the sort that were of central importance for Hermeticism: the curative power of amulets and weapon-salves; stellar virtues; the Alkahest; and transmutation" (Rattansi, "Origins," 134–135 and 139). Birch mentions Boyle's interest in "Butler's stone, the sympathetic powder, the weapon salve, the alkahest [universal solvent] of *Paracelsus*, the *virgula divina* [divining rod], the transmutation of metals, projection, or the philosopher's stone" (*W*, I, cxlvii).

Boyle was certainly familiar with the *Corpus Hermeticum*, including the *Asclepius*, that part of the *Corpus* setting out magical practices, for he quotes from it in support of his argument that the study of nature is good and should be encouraged because, among other things, it leads one to appreciate God's creation and worship him accordingly. Discovering the perfections of God in his creatures is, Boyle argues, "a more acceptable act of religion, than that burning of sacrifices. . . . *Trismegistus*, forbidding *Asclepius* to burn incense, tells him, 'That the thanks and praises of men, are the noblest incense that can be offered up to God'" (*W*, II, 63; see also his approving quotation of "the great *Mercurius Trismegistus*" at II, 31).

Boyle agreed with the Hermeticists that the world is a temple and the natural philosopher is qualified as the priest to "offer up for the creatures the sacrifice of praise to the creator; for . . . reason is a natural dignity, and knowledge a prerogative, that can confer a priesthood without unction or imposition of hands" (*W*, II, 8). One of his strongest statements of this point is his quotation from Philo: "The whole world is to be accounted the chiefest temple of God; the *Sanctum Sanctorum* of it is the purest part of the universe, heaven; the ornaments, the stars; the priests, the ministers of his power, angels, and immaterial souls" (*W*, II, 32). Harold Fisch points out that his quotation of Philo is particularly significant because Philo's "doctrines are impressed strongly upon the *Hermetica* in particular his belief in the holiness of the Kosmos . . . [and he] was consciously drawing an analogy between the world of Nature and the world of Scriptural revelation—the Temple he has in mind being the Temple at Jerusalem with its priests and Levites."[2]

Michael Hunter has recently shown us that Boyle had a strong interest in and anxiety about magic. Both his published and unpublished work provide evidence that he maintained this interest through-out his life. Sometime during the last two years of his life, Boyle dictated five pages of notes to his friend Gilbert Burnet, and Hunter points out that half of this material discusses magical phenomena, including the transmutation of metals using the philosopher's stone, the conjuring of spirits, the use of magical looking glasses, and predictions of the future.[3] Two accounts in particular make it clear that Boyle not only believed transmutation to be possible, but believed that it had actually been achieved. According to Burnet's notes, Boyle was given a grain of something "pretended to be the powder of projection," which was supposed to cause transmutation. Unfortunately, the donor lost most of the powder when he burned his hand putting the powder into hot lead, but Boyle retrieved a microscopic fragment of it. He "put [it] to lead in fusion which being so litle could not have great Operation but it stopt the fusion of the lead which made him conclude that a greater quantity might have produced some more considerable change." That is, Boyle put the fragment into melting lead and the melting stopped but nothing else happened, leading Boyle to conclude that a greater quantity would have had "great Operation." In the second account, Boyle watched someone put a "certain amount of 'a bright powder' into a crucible with a certain amount of lead." When the crucible cooled, Boyle was "not a litle surprised to find instead of lead Gold which after all the trialls that could be made was found true Gold" (Hunter, "Alchemy," 388–389).

Hunter reminds us that transmutation seemed feasible under the corpuscular philosophy, and "Boyle seems to have carried out transmutation experiments in the 1650s, which he referred to allusively in his publications of the 1660s, particularly in his *Origine of Formes and Qualities* of 1666[, wherein] he actually stated that the doctrine that he there expounded made 'the alchymists hopes of turning other metals into gold appear less wild.'" Boyle also wrote (in the 1670s, Hunter suggests) a dialogue on "the Generation and Transmutation of Metals," answering several objections to the possibility of transmutation and recording experimental findings "which supported the validity of transmutation, some of them evidently his own" (Hunter, "Alchemy," 399–401). He published a part of this dialogue in 1678, giving a detailed report of an experiment in which he used a small amount of dark red powder (a gift from a stranger) and found that it changed gold into silver (*W*, IV, 371–379).

Boyle evidently saw a clear link between the philosopher's stone and intercourse with spirits, for, in many of the accounts he gives, those conversing with spirits expect the spirits to reveal "insights into secrets concerning the philosopher's stone." For example, Burnet's notes tell us, "a man of quality and Estate" and "a Fellow of the Royal Society" to whom Boyle "gave a much more undoubted credit" saw the use of a magical glass to find objects stolen from him. Since only virgins could see anything in magical glasses, the gentleman brought to the priest who had the glass a nine-year old girl who described what she saw: his pockets being picked by a servant while asleep aboard a boat from Padua to Venice. He later brought "a processe of the Philosopher's stone . . . and made the priest ask if it was a good one." (The spirits became angry and refused to say.)

Boyle also began a dialogue discussing the pros and cons of using the philosopher's stone to conjure spirits. In it, Parisinus, a man of ample means, "versd in rationall and experimentall Phylosophy," says he wants to use the philosopher's stone to talk with good spirits. Interlocutors object that it is difficult to tell good spirits from bad, that undue attention to good angels might be idolatrous, and that "the ambition to attract angels by turning silver into gold was 'fitter for some whimsicall Enthusiast then for so learned a man as Parisinus.'" Parisinus responds that there is good reason to believe intercourse with spirits is possible and cites examples from the scriptures and other sources. The dialogue is incomplete, but the manuscript contains a list of further arguments: it is likely that "aire and fluid parts of the world" contain spirits, it is quite easy to tell good spirits from bad and to avoid worshipping good ones, and "there may be congruities or magnetisms capable of inviting them which we know nothing of" (BP 7, fols. 134v–135, 136–150). The dialogue pictures alchemy, Hunter notes, "as a route to communicate with spirits, presumably along the lines indicated by Elias Ashmole," who thought that "the highest aim of the alchemist is to achieve 'the *Angelicall Stone*,' which 'affords the *Apparition* of *Angells*, and gives a power of conversing with them, by *Dreames* and *Revelations*.'"[4]

Boyle also intended to publish a collection of accounts of supernatural phenomena such as second sight and apparitions in a magical looking glass. His goal was probably to defeat atheism, for, as Boyle wrote in a letter, "any one relation of a supernatural phaenomenon being fully proved, and duly verified, suffices to evince the thing contended for; and, consequently, to invalidate some of the atheists plausiblest arguments" (*W,* VI, 58). Hunter suggests that Boyle also aimed "to vindicate the mechanical philosophy from the accusation that it was itself dangerously associated with the spread of materialism and atheism" (Hunter, "Alchemy," 395).

And finally, Boyle wrote or planned to write several "Tracts Relating to the Hermetical Philosophy";[5] and, he wrote to a friend,

> since I find myself now grown old, I think it time to comply
> with my former intentions to leave a kind of Hermetic legacy
> to the studious disciples of that art, and to deliver candidly, in
> the annexed paper, some processes chemical and medicinal,
> that are . . . more of kin to the noblest Hermetic secrets, or, as
> *Helmont* stiles them, *arcana majora.* Some of these I have made
> and tried; others I have . . . obtained, by exchange or other-
> wise, from those that affirm they knew them to be real, and
> were themselves competent judges, as being some of them
> disciples of true adepts, or otherwise admitted to their acquain-
> tance and conversation. (*W,* I, cxxxi)

Marie Boas has claimed that Boyle's "addiction to a belief in transmutation arose from his staunch belief in the corpuscularian philosophy" and "while misguided, does not mean that Boyle was slipping into mystical alchemism." But from this evidence, we may surmise with Hunter that "that was indeed precisely what he was doing."[6] We are not surprised by Boyle's reference to his chemical "processes" as Helmontian *arcana majora,* for there is good evidence that he seriously considered Helmont's view that water is the basic element and that he adopted suitably mechanized versions of agents such as spirits, seminal principles, and ferments to explain motion and change in many substances and in animals, including humans.

Boas dates an early version of *The Sceptical Chemist* to no earlier than 1651 and no later than 1657.[7] In this manuscript, *Reflexions on the Experiments vulgarly alledged to evince the 4 Peripatetique Elements, or ye 3 Chymicall Principles of Mixt Bodies,* Boyle criticizes arguments adduced

for Aristotle's four elements and for Paracelsus's three principles. As
Boas notes, however, he is almost persuaded by Helmont's view. Boyle
considers three of Helmont's arguments in favor of water as a basic
principle of matter, including the resolution of things into water (by
strong "menstrums" or an alkahest, i.e., solvents), and remarks, "I
have not without some wonder observed in the analysis of bodies,
how great a share of water goes to the making up of divers [things]."
Of these, he discusses wood, eels, human blood, and corrosive spirits
(Boas, "Early Version," 165).

In support of Helmont's appeal to various experiments to prove
that bodies are "nothing but water, subdued by seminal vertues,"
Boyle singles out "this considerable Experiment": Helmont dried and
weighed some earth (200 pounds), planted a five-pound sapling in it,
and gave it only rainwater or distilled water. Five years later he dug it
up, dried and weighed the earth again, and found only three ounces
missing; yet the tree then weighed more than 169 pounds. Helmont
concluded "that 164 lb of the roots, wood and bark, which constituted the tree, sprung solely from the water" (Boas, "Early Version,"
166). Boyle describes his own attempts to perform this and similar
experiments and says,

> I chose springwater, rather then raine water, because the latter
> is more discernibly a kind of πανσωερμα containing in it (be-
> sides the celestial influences or steams of the heavenly bodies,
> which are supposed to impregnate it) a considerable and fertil-
> izing Earth and salt, which may be extracted out of it, and is
> by some mistaken for the spirit of the world corporified. I have
> had in my thoughts to make some trials, how experiments to
> the same purpose with Helmonts would succeed in other, then
> vegetables: but I have been hitherto hindred from it. Only I
> must admire the strange power of the formative power of the
> seeds of things, which doe not only fashion the obsequious
> matter according to the exigency of their owne natures, and
> the parts they are to act; but doe also so dispose and change
> the matter, they subdue, as to give it a consistency, which it
> seemed incapable of admitting, or we may observe in Eggs,
> where the seminal particles, tho at first scarce discernible to
> the Eye . . . doe not only prepare the matter into that great va-

riety of contextures and Consistencies, that is requisite to the
production of veine, sinew, artery, tendon, flesh, membrane,
gristle, the aqueous, vitreous, and chrystalline humors of the
Eye . . . but dos also out of the same matter produce the bones,
so much harder then that soft and liquid substance, whence
they are made. . . . The like we may observe in the sap of trees,
which the seminal vertue diffused in the twig of a grafted
peach or apricock, is part of it, hardned into that resisting substance, we cal the stones. And this induration of the sap of
trees I have yet more observed in the Indian cacaonuts where
tho the midle part of the kernel contain a liquor copious
enough . . . yet the shell of it is of such a hardnes and tough-
nes, that they endure polishing. . . . I could adde, that we observe in divers subterraneous caverns, that the water, that is
just ready to fall to the ground, is intercepted in that posture,
and by the petrifique seed or spirit that liquid substance grows
solid a veue d'oeil. . . . But let this be enough at present. So
that Helmonts opinion is worthe the considering; though not
presently of believing, till he bring also experiments of the production of a metal or mineral out of water. (Boas, "Early Version," 167)

The "strange power of the formative power of the seeds of things"
which Boyle admires is semina, a central concept for Helmont and
other natural magicians (and dating back at least to the Stoics). In
this passage, Boyle admires the ability of the semina to take one basic
element, Helmont's "insipid water," and change it into such different
things as soft flesh and hard bones, or into both the milk of the co-
conut and its hard shell. And by the "petrifique seed or spirit," water
in caves grows solid before our very eyes. Until he saw minerals produced from water, Boyle hesitated to believe that everything is made
from it, but he did not hesitate to endorse Helmont's seminal theory,
including the understanding of semina as spiritual agents (see his reference to "the petrifique seed *or spirit*"). We also note that he takes
for granted the existence of celestial influences and worries that they
make the use of rainwater in his experiment problematic.

While it is clear that Boyle never gave up his lively interest in Helmont's theories or his belief in seminal principles, ferments, and spir-

its, there has been some controversy over just how non-mechanical these agents were for him. John Henry has pointed out that many English natural philosophers found Descartes's strict mechanism untenable. In the first place, they thought that completely passive matter and a system in which all bodies are moved by bodies already in motion does not allow for the generation of new motion in the world and so could not adequately account for the data. And in the second, strict mechanism supports atheism on the one hand and dangerous heterodoxies on the other. As Henry notes, "If the world system was so contrived that once God set it in motion it could continue indefinitely without his further intervention, then it was easy to imagine that God was not necessary at all."[8] Thus, Boyle and others had recourse to active principles.

For example, in "Suspicions about Some Hidden Qualities in the Air," Boyle tells us,

> The difficulty we find of keeping flame and fire alive, though but for a little time, without air, makes me sometimes prone to suspect, that there may be dispersed through the rest of the atmosphere some odd substance, either of a solar, or astral, or some other exotic nature, on whose account the air is so necessary to the subsistence of flame; which necessity I have found to be greater, and less dependent upon the manifest attributes of the air, than naturalists seem to have observed. [Here Boyle describes experiments like Helmont's in which small burning lamps are enclosed in glass containers and the flames expire very quickly.] For after the extinction of the flame, the air in the receiver was not visibly altered, and, for aught I could perceive . . . the air retained either all, or at least far the greatest part of its elasticity, which I take to be its most genuine and distinguishing property.
>
> And this undestroyed springiness of the air seems to make the necessity of fresh air to the life of hot animals . . . suggest a great suspicion of some vital substance, if I may so call it, diffused through the air, whether it be a volatile nitre, or (rather) some yet anonimous substance, sydereal or subterraneal, but not improbably of kin to that, which I lately noted to be so necessary to the maintenance of other flames. (W, IV, 90–91)

Flames do not depend on the manifest attributes of the air as much as naturalists have thought. Experimental evidence suggests that there is "some vital substance" of a "solar, or astral, or some other exotic nature" in the air that allows flames to burn and which is necessary to the life of hot animals.

What marks this account is that, although Boyle suggests that there is probably some "odd," "vital substance" in the air, functioning like Helmont's *magnale*, he has dropped the animism so characteristic of the natural-magic tradition. Boyle's version of the *magnale* is not the World Spirit, not a "third thing" between body and soul.

In the same way, celestial magnets attract sidereal and celestial effluvia (things given off by the stars and planets), but not by sympathy. Having explained sympathy and antipathy as depending "on the peculiar textures and other modifications of the bodies, between whom these friendships and hostilities are said to be exercised," Boyle turns to explain celestial magnets; some bodies, he says,

> may be receptacles, if not also attractives, of the sidereall, and other exotic effluviums, that rove up and down in our air.
>
> Some of the mysterious writers about the philosophers-stone speak great things of the excellency of what they call their philosophical magnet, which, they seem to say, attracts and (in their phrase) corporifies the universal spirit, or (as some speak) the spirit of the world. But these things being abstrusities, which the writers of them professed to be written for, and to be understood only by the sons of art; I, who freely acknowledge I cannot clearly apprehend them, shall leave them in their own worth as I found them. (*W, IV*, 95–96)

Boyle does not tell us what the "celestial effluviums" are, but they are certainly not incorporeal spirits with perception, intelligence, and will. And he neither affirms nor denies the suggestion that the World Spirit is "corporified." (The picture here is of the thickening of a body made of ether, i.e., not quite incorporeal spirit, not quite substantial matter.)

Thus, while it might be said that Boyle has recourse to active principles, corpuscles that move without being pushed by something already in motion, he certainly eschews animism and the World Spirit. The seminal principles are active, but no longer contain archei;

Boyle's "vital substance" in the air is not Helmont's *magnale;* sympathy and antipathy are reduced to peculiar modifications of bodies; and other occult qualities are due to the "general fabrick" of the universe (*W,* IV, 86, 89, 95–96; III, 306–307, 316).

Henry suggests that Boyle held back "his real thoughts on the matter" of "unheeded agents" producing cosmic qualities and downplayed his use of active principles (Henry, "Cosmical Qualities," 133). Certainly when he rewrote the *Reflexions* for publication in 1661 as *The Sceptical Chemist,* Boyle's admiration for "the strange power of the formative power of the seeds of things" and the many examples of these semina changing "obsequious matter" into such a variety of hard and soft bodies disappear, replaced by reference to the many ancient sources and authorities who endorsed the view that water is fundamental (*W,* I, 497ff.). He also cleans up the alchemical language found in his discussion of rainwater; instead of containing "celestial influences or steams of the heavenly bodies, which are supposed to impregnate it," it contains "the steams of several bodies wandering in the air, which may be supposed to impregnate it" (*W,* I, 495; see also III, 69ff).

Henry argues that Boyle downplayed his use of active principles because of the political and religious meaning attributed to them by contemporaries: "whenever he could without compromising his natural philosophy, he wrote as though matter was entirely passive" (Henry, "Occult Qualities," 356). He points out that Boyle's contemporaries did not always distinguish between active matter (moving without a push from a body already in motion) and animate matter (matter with the ability to perceive, or with intelligence and will). For example, the Cambridge theologian Henry More was well aware of the dangers of "enthusiastic" beliefs in animate matter; one of the two beliefs More identified as enthusiastic was

> That everything has *Sense, Imagination,* and a *fiducial Knowledge*
> of God in it, *Metalls, Meteors* and *Plants* not excepted.

The danger of enthusiasm ultimately caused More to eschew not only animate matter, but even active matter. Nevertheless, despite this danger, Boyle was moved to accept active matter because it was necessary to explain some phenomena (and because of his theological concerns).[9]

My own view is that Boyle downplayed his use of active principles because of the danger of Epicurean atheism, on one side, and subversive enthusiasm, on the other. Epicureans argued that matter could move itself (so there was no need for a deity to set things in motion) and Enthusiasts argued that matter could move itself because it was animated or inspirited and had knowledge of God, and used this belief as a foundation for their political aspirations. And although Boyle made use of active principles because he found it difficult to explain certain phenomena, especially in biology, without them, there is no clear evidence that he was ever committed to Hermeticism.

Thus, although he believed in magic, in such alchemical agents as the philosopher's stone, the powder of projection, and transmutation, in celestial effluvia, seminal principles, ferments, and other "odd substances," Boyle pursued these matters within the mechanical philosophy, not within a hylozoic one, for he rejected the animation of matter and the World Spirit. The question is, why did he pursue a mechanistic research program?

In part, of course, because in Boyle's words, the mechanical or corpuscular philosophy "comports well" with the data. In chapter 17, we will examine the claim that his decision to pursue research within the mechanical paradigm was driven solely by the data. Next, however, we will examine the evidence for our claim that he rejected animism and the World Spirit and pursued his interests within a mechanistic paradigm in part because animism was associated with radical enthusiasm.

11

Hermeticism, Hylozoism, and Radical Politics

Peter Rattansi and others have argued that early on in his life Boyle was committed to Hermeticism, but that he switched to the mechanical philosophy in the mid-1650s because he wished to distance himself from the radicals whose political projects rested upon Hermetic foundations.[1] This claim is the subject of much controversy.

Henry agrees that "various Paracelsian/Helmontian world views were associated by contemporaries with subversive 'enthusiasm' in religion," but denies that because of this association, Boyle and others rejected active matter and switched to mechanism, with its view that matter is passive.[2] Clericuzio also argues that Boyle believed active corpuscles constitute various non-mechanical agents such as spirits, seminal principles, and ferments, and that he used them to explain motion and change in many substances and in animals, including humans. But because he maintained his interest in these active, non-mechanical agents throughout his life, Clericuzio says, "it would be very difficult to state that Boyle's moving to Oxford entailed his abandonment of his juvenile Helmontian enthusiasm and his adhesion to a sound mechanical philosophy."[3] Let us take it as given that fear of their association with subversive enthusiasm did not lead Boyle to reject active principles and embrace strict mechanism. Boyle, along with other moderate mechanists, was willing to negotiate his way through the dangers from both the "left" and the "right" on the issue of active principles. But he clearly rejected animism, and there is evidence that Boyle rejected the hylozoic experimental research program because of the association between subversive enthusiasm and animism.

Hunter argues that, although Boyle was very interested in magic, he did not pursue it because he feared the moral evil that might follow from its practice. (He seems to have been afraid, among other things, that the conjuror of evil spirits might become subject to them, and that a discoverer of the Elixir of Life might become dissolute.) Undoubtedly Boyle worried about the moral implications of magic; however, Hunter also tells us that

> in a paper that Boyle wrote about one 'Process,' he noted that it had the advantage that 'this work is not of that benefit as may threaten the welfare of States, if it should fall into unworthy hands.' Another dialogue consciously limited itself to those aspects of the validity or otherwise of the philosopher's stone which eschewed 'moral or Political Considerations,' while in a paper about the pros and cons of revealing the 'great Arcanum,' one of the reasons against it was the risk 'That it would much disorder the affairs of Mankind, Favour Tyranny, and bring a general Confusion, turning the World topsy turvy.'[4]

Political worries are clearly expressed in these quotations. Indeed, "turning the world topsy turvy" was a standard way to refer to the revolution that the radicals were blamed for threatening to bring about, or for having caused. We can see that Boyle was caught between his own interests in Hermetic doctrines and practices and his fear, first, that his alchemical work would fall into the wrong hands and be used to support radical politics, and, second, that he would himself be accused of aiding and abetting radicals in this way. In his *Philosophical Transactions* article of 1676, as Hunter notes, "he inquired whether potential medicinal uses of his mercury were likely 'to exceed the *political* inconveniencies, that may ensue, if it should prove to be of the best kind, and fall in ill hands'" (Hunter, "Alchemy," 407, quoting *W,* IV, 227–228, emphasis mine).

Again and again throughout his scientific work, Boyle says that his mechanistic hypotheses "accord well" with the data, but he also eschews the "morals and politicks" associated with animism. He maintains respect for Helmont, but objects to Paracelsus's and Helmont's chemical hypotheses that ascribe feelings such as sympathy and antipathy to acids and alkalis, for, he says,

> Those hypotheses do not a little hinder the progress of human knowledge, that introduce morals and politicks into the explica-

tions of corporeal nature, where all is transacted according to laws mechanical. (*W,* IV, 291)

Hermeticism, and particularly its commitment to hylozoism, was associated with revolutionary hopes from the time of the European lay spiritual movements in the thirteenth century through the Elizabethan era in England, to the time of Boyle (and beyond). As the turmoils of the seventeenth century began, the association between the natural-magic tradition and political rebellion became clearer. In 1600, Campanella was tortured by the Inquisition for rebelling against Spanish rule in Naples in order to set up a "universal republic." Campanella describes the republic in his *City of the Sun,* published in 1623 while he was imprisoned: it is, as we have noted, led by a Hermetic magician and characterized by communal ownership of property. Campanella imagines an ideal human community living together in love. This vision mirrors his view of nature: the basic constituents of both the human and natural communities are credited with life and consciousness. Concerning the vacuum, for example, he writes:

> All bodies abhor the existence of a vacuum, and they rush, with natural impetus, to fill such a void in order to conserve the community entire. This is because all enjoy being together, and cherish their reciprocal contact with one another—which contact comprises their common life. Thus we see that air, in the depths of the sea or in the earth's cavities, will descend with great impetus in order to prohibit the formation of a vacuum; almost . . . seeming to express a particular hatred for the water or the earth into which it rushes for the purpose of promoting the common good.[5]

(Nevertheless, Campanella believes that a vacuum can be created by violence, offering such examples as sealing a bellows with pitch and forcing it to rise, "the cupping-glass pull[ing] at the flesh because it is empty of air," and others (*Sense and Feeling,* 364).

In seventeenth-century England, revolutionaries of many sorts held a natural philosophy that grew out of certain theological heresies in turn derived from the Hermetic natural-magic tradition. The preeminent Digger Gerrard Winstanley offers a clear case of the sectarian debt to the natural-magic tradition. He held a kind of materialist pantheism which identified God with the created world and so placed the spirit of life and cause of motion within terrestrial and celestial bodies themselves:

> To know the secrets of nature is to know the works of God. . . .
> And indeed if you would know spiritual things, it is to know
> how the spirit or power of wisdom and life, causing motion or
> growth, dwells within and governs both the several bodies of
> the stars and planets in the heavens above; and the several bod-
> ies of the earth below, as grass, plants, fishes, beasts, birds and
> mankind.[6]

And Christopher Hill tells us that, like Winstanley, many of the Ranters adopted a materialist pantheism. Lawrence Clarkson, who served in Cromwell's New Model Army, also combined radical economic, social, and religious (including natural-magic) views. As a Ranter, he believed that "God was in all living things and in all matter," and in 1647 he asked, "Who are the oppressors but the nobility and gentry . . . and who are oppressed, if not the yeoman, the farmer, the tradesman and the like? . . . Your slavery is their liberty, your poverty is their prosperity." Clarkson also spent a period as an astrologer and magician (Hill, *Upside Down,* 206, 214, 217).

Jacob Bauthumley was still in the army when, in 1650, he wrote that "[a]ll the creatures in the world . . . are but one entire being" and there is "[n]othing that partakes of the divine nature, or is of God, but is God." And finally:

> Not the least flower or herb in the field but there is the divine
> being by which it is that which it is; and as that departs out of
> it, so it comes to nothing, and so it is today clothed by God,
> and tomorrow cast into the oven. (Hill, *Upside Down,* 219)

For publishing the work from which these remarks are taken, Bauthumley was bored through the tongue, a common punishment for those whose work was condemned as blasphemous.

John Pordage expressed this view in somewhat more sophisticated language, as befitted "a gentleman and student of All Souls in Oxford." Hill tells us that

In Pordage himself "that inward spiritual eye, which hath been locked up and shut by the Fall," was "opened in an extraordinary way." It revealed to him that "there were two invisible principles . . . two spiritual worlds extending and penetrating throughout this whole visible creation."

Pordage combined radical class with radical gender politics. In 1655 he was accused not only of Ranter views but also of saying that "there would soon be no Parliament, magistrate or government in England; that the saints would take over the estates of the wicked"; and Richard Baxter wrote that Pordage was "much against property, and against [hierarchical] relations of magistrates, subjects, husbands, wives, masters, servants, etc." Pordage approved of Richard Coppin, who, in his 1649 *Divine Teachings,* wrote that "God is all in one, and so is in everyone." And he was thought to follow Thomas Tany, who denied that anyone can lose salvation, denied that hell exists, and believed that God is in everything (Hill, *Upside Down,* 225–226).

Finally, for the Quakers, the principle that the spirit of life is in all creatures was expressed in the doctrine of the inner light, according to which the spirit within or the spirit of God within reveals God's will and reveals the truth to the individual. These beliefs about nature and about the inner light harmonized perfectly with a revolutionary commitment to human equality, for all people have (or can have) God-within, all can know God's will, not just priests and bishops. Sectaries asked the next questions: What need, then, for a hierarchical state church? Why should people of God pay tithes to support a church whose doctrines they disagree with? Nor, as we have seen, would such a revolution be limited to the church, since, as James I remarked, "No bishop, no king." The state and the church, as he recognized, depended upon one another.

When sectaries turned their attention to reformation of the universities, they hoped that thinkers from the natural-magic tradition would replace Aristotle. John Webster plumped for the "philosophy of *Hermes,* revived by the Paracelsian School," and hoped that university students would become careful experimenters:

not idely trained up in notions, speculations, and verbal disputes, but may learn to inure their hands to labour, and put their fingers to the furnaces, that the mysteries discovered by

Pyrotechny, and the wonders brought to light by *Chemistry,* may
be rendered familiar unto them . . . ; that so they may not be
Sophisters, and *Philosophers,* but *Sophists* indeed, true Natural Magicians . . . in the center of nature's hidden secrets, which can
never come to pass, unless they have Laboratories as well as Libraries. (Quoted in Rattansi, "Paracelsus," 27)

But sectaries were not the only ones who adopted views derived
from the natural-magic tradition. Rattansi notes that both Paracelsian
and Helmontian doctrine "exalted the knowledge of *illumination* above
that derived from 'carnal reason'" and that this mystical, anti-rational
doctrine appealed to many Englishmen during the civil crisis in England (Rattansi, "Paracelsus," 26). For example, Walter Charleton
was a royalist, appointed physician to Charles I and later a Fellow of
the Royal Society. In 1650, he published the first English translation
of Helmont's work and declared for his doctrines. He stressed particularly that reason is corrupt and the cause of the current religious
turmoil:

we must quit the dark Lanthorne of Reason, and wholly throw
ourselves upon the implicit conduct of faith. For a deplorable
truth it is, that the unconstant, variable, and seductive impos-
ture of Reason, hath been the onely unhappy Cause, to which
Religion doth owe all those wide, irreconcilable and numerous
rents and Schisms.[7]

Rattansi points out that, not long afterward, Charleton recanted his
adherence to Helmont in favor of mechanism. Helmont had explained
the weapon-salve (which cured wounds by its application, not to the
wound, but to the weapon that caused it) in terms drawn from natural magic. In 1654, Charleton dismissed the cure as ridiculous and
Helmont as "Hairbrain'd and Contentious."[8] Between 1650 and 1654,
Charleton expressed concern at the sectarian threat to religion and
society, rooted in the sectaries' belief that they could know God's will,
and argued that Descartes offered a way to refute that belief.

The natural-magic tradition had come to be identified, "through
the sectarians and their 'inner light' doctrines, with heretic religious
and social opinions." Thus, Thomas Hall could attack John Webster
by arguing that his belief in private illumination and in the doctrines

of Paracelsus demonstrated "that he belonged to the 'Familiasticall-Levelling-Magicall temper'" (Rattansi, "Paracelsus," 31 and 29). Here the natural-magic tradition ("Magicall") is clearly identified with subversive politics ("Levelling") and religion ("Familiasticall" refers to the sectarian Family of Love).

12

Boyle's Concern over the Sectaries

Boyle agreed with the sectaries that "the right, and the skill to govern, are two very distinct things: nor does the one confer the other." There certainly are rulers whose "power is unguided by prudence," but Boyle was not sympathetic with the democratic political aspirations of "the vulgar." There must needs be a ruler and he must be obeyed even when he is not wise, for

> the vulgar, who yet make up the far greatest and loudest part of those, that would intrude themselves into state-affairs, upon the pretence of their being ill managed by their superiours . . . whatever the course of affairs be, these cannot but be incompetent judges of their being politick. . . . the vulgar is rarely admitted to have such a prospect into the true state of affairs, as is requisite to enable them to judge of the expedience or unadvisedness of them.[1] (*W,* II, 412–413)

A careful reading of Boyle's writings from this period reveals his preoccupation with the sectaries and his determination to undercut the natural philosophy upon which their religious and social doctrines rested. Thus, *Some Considerations Touching the Style of the Holy Scriptures,* written in 1651 and 1652, attacks certain sectaries; it proceeds by defending the scriptures against detractors, and though he states that it is not meant to deal with "atheists and antiscripturalists," it nevertheless does so in many places. The term "antiscripturalist" was used not precisely but rather pejoratively, to refer to a range of sectarian attitudes toward the scriptures. Roughly, these sectaries did not defer to the Bible as authoritative, nor did they defer to authorized readers of it; they denied that the scriptures provide the primary point of contact with God. Instead, the sectaries held, as we have seen, that God speaks directly to his people through the inner light, or his spirit within. For some, the spirit could reveal the true meaning of the scriptures; for others, the spirit might reveal the scriptures to be irrelevant. Potentially, then, each individual becomes an authority.

Hill provides numerous examples of sectaries who, as Boyle put it, "profaned" the scriptures and alarmed their fellows. Perhaps the best exemplar of the connections among sectarianism, natural magic, and anti-scripturalism is John Everard (1575–c. 1650), who was

> fined under Laud for Familism, Antinomianism and Anabaptism. Everard translated Hermes Trismegistus and many works of mystical theology, including 'that cursed book', *Theologia Germanica*. He thought God was in man and nature . . . and allegorized the Bible. "The dead letter is not the Word, but Christ is the Word," he said. "Sticking in the letter" has been "the bane of all growth in religion."[2]

(Boyle himself used Everard's translation of *Hermes Trismegistus;* see *W,* II, 57.) Many others were also charged with allegorizing the scriptures (Hill, *Upside Down,* 218ff.).

Boyle was very concerned about the threat these sectaries presented, remarking in the *Style,*

> Wherefore, as in infectious times, when the plague reigns . . .
> so now that anti-scripturism grows so rife, and spreads so fast,
> I hope it will not appear unseasonable to advise those, that tender the safety and serenity of their faith, to be more than ordinarily shy of being too venturous of any books, or company, that may derogate from their veneration of the scripture; because by the predominant and contagious profaneness of the times, the least injurious opinions harboured of it, are prone to degenerate into irreligion. But I fear, you will think I preach. (*W,* II, 294–295)

The best defense against these errors is studying the Bible to improve one's "reverence for the scripture it self, and Christianity in general" and to find "solid evidences of that great truth, that the scripture is the word of God, which is indeed the grand fundamental."[3]

We must continually bear in mind that for seventeenth-century Christians there was little distinction between proper religion and the good society. Boyle was typical in holding that the scriptures teach truths about government and society; the view was that, as Jacob remarks, "Good government, on the whole, is informed by scriptural religion" (Jacob, "Boyle's Circle," 131).

We know that Boyle kept abreast of sectarian threats, for among the Boyle Papers is a set of notes taken on sermons preached during the late summer and autumn of 1655 at Allhallows, London, by "leading Fifth Monarchy Men, including John Simpson and Cornet Day."[4] And Birch tells us that Boyle himself attended a sectarian conventicle at Sir Henry Vane's house and there heard Vane preach that "many doctrines of religion, that had long been dead and buried in the world, should before the end of it be awakened into life"—a worry to Boyle the decade before.[5] However, at this meeting, he took issue with Vane's allegorization of the passage from the book of Daniel he had taken as his text, and Birch says that "[w]hen Sir Henry had concluded his discourse, Mr. Boyle spoke to this effect to him before the people; . . . that this place in Daniel being the clearest one in all the Old Testament for the proof of the resurrection, we ought not to suffer the meaning of it to evaporate into allegory." Boyle remarked that he objected in order "that the sense of the scriptures might not be depraved" (W, I, cxl–cxli).

When we examine the ways in which Boyle's concern about the radicals intersected his science, we find that the sharp distinction we in the twentieth century make between science and religion renders it difficult for us to imagine the context in which Boyle lived and worked: one in which the truths of nature and the truths of religion were part of the same system. (In fact, Boyle gets some of the credit for creating the distinction we have become accustomed to.[6]) But given the background assumption that there is one system of truth, we can understand Boyle's belief that a study of scripture (God's words) and of nature (God's works) reveal the same truths:

> [God] hath been pleased to contrive the world so, that . . . it may afford [man] not only necessaries and delights, but instructions too. For each page in the great volume of nature is full of real hieroglyphicks, where (by an inverted way of expression) things stand for words, and their qualities for letters. . . .
>
> Nor can the creatures only inform man of God's being and attributes . . . but also instruct him in his own duties; for we may say of the world, as St. Austin did of the sacraments, that it is *verbum visibile* [the word made visible]. (W, II, 29)

Since "his own duties" refers to all duties, including not only religious and moral but also social and political ones, it follows that a study of nature will reveal proper moral, political, and social arrangements. The correct understanding of nature would therefore be well worth contesting, for that understanding yields the proper political and social order.

Boyle himself makes the connection between hylozoism and sectarianism in *A Free Enquiry,* the essay expressing his clearest and strongest arguments for the mechanical philosophy and "the death of nature" against its great hylozoic rival, "the vulgarly received notion of nature." (By "vulgarly received," Boyle means the commonly accepted view.) Here Boyle distinguishes among the meanings commonly ascribed to the term "nature" and teases out the one he is concerned to expunge from use: "most commonly, we would express by the word nature a semi-deity, or other strange kind of being, such as this discourse examines the notion of" (*W,* V, 167). There is no one clear doctrine of nature in this sense, of course, so Boyle is forced to respond to several related ideas derived ultimately from Aristotle and Plato; the "schools," that is, followers of Aristotle, "have been the chief propagators" of the "commonly received opinion of nature," but it is also held "by philosophers and other writers, and by the generality of men." According to this opinion,

> Nature is a most wise being, that does nothing in vain; does not miss of her ends; does always that, which . . . is best to be done; and this she does by the most direct or compendious ways, neither employing any things superfluous, nor being wanting in things necessary; she teaches and inclines every one of her works to preserve itself: and, as in the microcosm, (man) it is she, that is the curer of diseases; so in the macrocosm (the world) for the conservation of the universe, she abhors a vacuum, making particular bodies act contrary to their own inclinations and interests, to prevent it, for the public good. (*W,* V, 174)

When he comes to explain why he objects to the opinion that nature is a "goddess, or at least a semi-deity," Boyle tells his readers that "the vulgar notion of nature has had, and therefore possibly may have," evil effects on religion. These effects are polytheism and idolatry, and they arise from "looking upon meerly corporeal, and oftentimes inanimate things, as if they were endowed with life, sense, and understanding." Although this view was evident in ancient times, he says, it is current today. Even now

> there is lately sprung up a sect of men, as well professing Christianity, as pretending to philosophy, who (if I be not misinformed of their doctrine) do very much symbolize with the antient Heathens, and talk much indeed of God, but mean such a one, as is not really distinct from the animated and intelligent universe; but is, on that account, very differing from the true God, that we Christians believe and worship. And though I find the leaders of this sect to be looked upon by some more witty than knowing men, as the discoverers of unheard of mysteries in physics and natural theology; yet their hypothesis does not at all appear to me to be new. (*W,* V, 183)

The opinions Boyle is concerned about, then, are a mix of pantheism and, literally, the inspiration or infusing of spirit into the world. These were precisely the two beliefs that Henry More decried in his discourse against enthusiasm:

> That Nature is the body of God, nay God the Father, who is also the World, and whatsoever is any way sensible or perceptible.

And

> That everything has *Sense, Imagination,* and a *fiducial Knowledge* of God in it, *Metalls, Meteors* and *Plants* not excepted.[7]

Boyle perceives that Nature corresponds to the World Spirit; there is, he notes, a "great affinity between the *soul of the world,* so much talked of among the heathen philosophers, and the thing, that men call nature." This affinity leads him to fear the resurgence of "the great and pernicious errors" that the old sages were led into

> by the belief that the universe itself, and many of its nobler parts, besides men, were endowed, not only with life, but understanding and providence. (*W,* V, 183, 184–185)

Boyle says the belief that everything has understanding and providence is dangerous: "it is a dangerous thing to believe other creatures, than angels and men, to be intelligent and rational." Not because Christians would then, "like some Heathens, worship nature as a Goddess," nor because it would "subvert, nor much endanger any principle of religion." Rather, there are "extravagant and sacrilegious errors" that have been embraced "by divers modern professors of Christianity, who have of late revived, under new names and dresses, the impious errors of the Gentiles." These people are "some of our late infidels," who have pretended "to be great discoverers of new light in this affair" (*W,* V, 188 and 250–251).

In 1666, when *A Free Enquiry* was written, Boyle still had reason to worry about "our late infidels," the radical sectaries, for they were active in the 1660s; his brother Roger, now the Earl of Orrery, reported in 1663, 1665, and early in 1666 "that there were sectarian plots on foot for rebellions in England and Ireland." Rumors of Fifth Monarchist uprisings were continuous after the Restoration despite the imprisonment of rank-and-file believers and the execution of their leaders and supporters (Rogers, *Fifth Monarchy Men,* 123–133). Too, sectaries eagerly seized the opportunity opened to them in 1665 and 1666 when the Black Death raged in London, sending the court to Oxford and everyone who had the means and the desire, including the clergy, out of the city. The sectaries stepped into vacant pulpits and preached sedition.[8] The sects and political radicals had been merely suppressed after the Restoration, not destroyed.

Boyle says these beliefs themselves, and not just those who hold them, are dangerous, because they can seduce those who are "wont or are inclined to have an excessive veneration for what they call nature . . . into those extravagant and sacrilegious errors" (*W,* V, 250). Thus, anyone who holds such views is suspect. You did not actually have to be a sectary to be accused of being one, or of being a fellow traveler, or of offering aid and comfort to sectaries. As sectarian hopes for greater democracy dimmed, first at the loss of the Republic and the change to the Protectorate and then at the restoration of the monarchy under Charles II, radical sectaries were blamed for the Civil War and for the regicide and were perceived as a great threat to the status quo. A cottage industry grew up around what we might call "radical baiting," which led to corresponding efforts to peel the radical label off one's own work and stick it on someone else's. Since the sectaries were often referred to as "enthusiasts," the charge of "enthusiasm" or of harboring "enthusiastic" principles was not to be taken lightly. For example, in 1666, Samuel Parker, an Anglican priest, blasted a group of courtiers claiming to be brethren of the Rosy Cross (members of an amorphous Hermetic group who proclaimed a reformation of all learning and derived many of their beliefs from Ficino's writings on natural magic) for enthusiasm and sedition:

> they directly Poison mens minds and dispose them to the wild-
> est and most *Enthusiastick Fanaticisme;* for there is so much
> Affinity between *Rosi-Crucianism* and Enthusiasme, that who-
> ever entertains the one, he may upon the same reason embrace
> the other. . . . And what Pestilential Influences the Genius of
> *Enthusiasme* or opinionative Zeal has upon the Publick Peace, is
> so evident from Experience [i.e., the Civil War], that it needs
> not be prov'd from Reason.[9]

The difference between enthusiasm and atheism seems straightfor-ward enough; from a moderate point of view such as Boyle's, the lat-ter posited no spirit while the former posited too much. However, the two were sometimes run together in the seventeenth century. Henry More's attack on enthusiasm, for example, was based on the premise that "Atheism and Enthusiasm, though they seem so extremely op-posite one to another, yet in many things do nearly agree."[10] But this can lead to historiographical problems, since it is not at all clear that atheism was actually associated with demands for radical social change in the way that enthusiasm was.

Michael Hunter has explored uses of the term "atheism" in the later seventeenth century and suggests several trends in society that accu-sations of atheism might have described, e.g., "naturalistic and secu-larist explanation" and "stress on reason and on natural causes; the retreat from providentialism and the miraculous."[11] But he is careful to distinguish atheism from enthusiasm:

> Not all of the disquieting features of contemporary life were
> included in the atheist stereotype. The attack on atheism thus
> remained separate from the attack on enthusiasm. ("Hetero-
> doxy," 448)

Hunter points out that there is little evidence to support a link between atheism and radicalism. Anti-atheists were primarily attacking "the general disrespect for accepted ideas and mores," particularly as expressed in fashionable circles, rather than "specifically political radicalism." And men like Boyle were also defending themselves from charges of atheism to which their mechanism made them susceptible. Thus, it would be wrong to see anxiety about atheism "as closely tied to any specific threat of radical change" ("Heterodoxy," 449, 456); Hunter particularly cites the Jacobs as making this mistake. Nevertheless, although More and others were not always careful to distinguish sectarian enthusiasts from atheists, moderate Englishmen like Boyle were certainly worried about "enthusiastic" radicals. The close tie between enthusiasm and specific threats of radical change was clear to Boyle and others at the time.

13

Boyle's Objections to Hylozoism

In this chapter, we will see that Boyle's objections to animism and a World Spirit were an important part of his objection to hylozoic research programs. As we have seen, *A Free Enquiry* has the effect of showing that the vulgar hylozoic philosophy gives more aid and comfort to enthusiasts than does the corpuscular philosophy. But, according to the corpuscular philosophy, Boyle tells us, there is no other "efficient" or cause of motion in the universe than God. There is no World Soul or Nature understood as a semi-deity inspiriting or animating matter. The world is just a "compounded machine" or "pregnant automaton, that, like a woman with twins in her womb, or a ship furnished with pumps, ordnance, etc. is such an engine, as comprises or consists of several lesser engines." Or again, "according to us, [the world] is like a rarer clock, such as may be that at Strasburgh, where all things are so skilfuly contrived, that the engine being once set a moving, all things proceed, according to the artificer's first design." As against the "extravagant and sacrilegious errors" that follow from the vulgar view, Boyle is quite clear that "our doctrine [the corpuscular philosophy] may keep many, that were wont, or are inclined to have an excessive veneration for what they call nature, from running, or being seduced, into those extravagant and sacrilegious errors" (*W, V*, 179, 163, 250).

A Free Enquiry tells us why Boyle objected to the notion of a World Spirit and the animism associated with it, and it sets up mechanism as a satisfying alternative theory. Earlier, in *Some Considerations Touching the Usefulness of Experimental Philosophy*, Boyle mentioned as his first objection to hylozoic accounts of nature the fact that they assume animism and a World Spirit.[1] The corpuscular philosophy is preferable, he says, precisely because it rejects them. He objects to "sympathy, antipathy and occult qualities" as explanations of magnetism and gravity (even though he admits that he has no alternative explanation). As an example of what it is he objects to about hylozoic accounts he discusses the suction of water through a reed: if an Aristotelian sucks liquid into his mouth through a reed, "he will readily tell you, that the suction drawing the air out of the cavity of the reed, the water must necessarily succeed in the place deserted by the air, to prevent a vacuity abhorred by nature" (*W*, II, 36, 37). The problem with this account, Boyle says, is that it assumes a World Spirit and that matter has perception, knowledge, and desire:

> to say, that the ascent of the water . . . proceeds from nature's detestation of a vacuity, supposes that there is a kind of *anima mundi*, furnished with various passions, which watchfully provides for the safety of the universe; or that a brute and inanimate creature, as water, not only has a power to move its heavy body upwards, contrary (to speak their language) to the tendency of its particular nature, but knows both that air has been sucked out of the reed, and that unless it succeed the attracted air, there will follow a vacuum; and that this water is withal so generous, as by ascending, to act contrary to its particular inclination for the general good of the universe, like a noble patriot, that sacrifices his private interests to the publick ones of his country. (*W*, II, 37–38)

By this account, Boyle complains, water is very conscious, indeed; it has general inclinations (e.g., to move downward to its proper place above the earth and below the air) which it will, reasoning from what it knows, generously sacrifice for the good of the community.

Introducing the mechanistic philosophy, he explains that the suction of water and other natural phenomena may appear to result from matter's desires, knowledge, and reason, but that is because God made them to follow his laws in the way that machines such as clocks perform orderly motions. For most phenomena of nature, except the intellect and will of man and possibly the actions of some other animals, Boyle says,

> we may, without absurdity, conceive, that God . . . having resolved, before the creation, to make such a world as this of ours, did divide . . . that matter, which he had provided, into innumerable multitude of very variously figured corpuscles, and both connected those particles into such textures or particular bodies, and placed them in such situations, and put them into

such motions, that by the assistance of his ordinary preserving
concourse, the phaenomena, which he intended should appear
in the universe, must as orderly follow, and be exhibited by the
bodies necessarily acting according to those impressions or
laws, though they understand them not at all, as if each of
those creatures had a design of self-preservation, and were furnished with knowledge and industry to prosecute it; and as if
there were diffused through the universe an intelligent being
[a World Soul or *anima mundi*] watchful over the publick good
of it, and careful to administer all things wisely for the good of
the particular parts of it, but so far forth as is consistent with
the good of the whole, and the preservation of the primitive
and catholick laws established by the supreme cause; as in the
formerly mentioned clock of Strasburg, the several pieces making up that curious engine are so framed and adapted, and are
put into such a motion, that though the numerous wheels, and
other parts of it, move several ways, and that without any
thing either of knowledge or design; yet each performs its part
in order to the various end, for which it was contrived, as regu-
larly and uniformly as if it knew and were concerned to do its
duty. (*W,* II, 39)

According to the corpuscular philosophy, God created matter, divided
it into atoms, set up the laws of nature, connected the atoms into bodies, and put them into lawful motion, and in this way caused the
world we know to come about. In themselves the bodies have no sen-
sations, knowledge, or intentions, though the laws of nature are so
well thought out by God that the atoms behave as if they were conscious and as if they followed the wise guidance of a World Soul.

We find, then, that the earliest arguments in support of the corpuscular philosophy insist upon the rejection of hylozoic accounts of nature because they assume animism and a World Spirit. And the latter
are objectionable because they can lead the unwary into heresies and
thence to behave like "our late infidels."[2]

Rose-Mary Sargent claims, to the contrary, that Boyle's natural theology "provided" his mechanistic ontology, and she dismisses earlier
attempts (e.g., by the Jacobs, Shapin and Schaffer, and others) to
show how Boyle's views of enthusiasm, atheism, and the debates over

social structures played a part in his choice of epistemological and metaphysical positions.[3] The implication of her claim that Boyle's natural theology provided his ontology is that he developed his theology of nature, including his account of creation, first and that it gave rise to a mechanistic ontology. She presses this claim in chapter 4 primarily through an examination of his inquiry into uses of the term "nature" in his work *A Free Inquiry into the Vulgarly Received Notion of Nature*.

In this work, Boyle gives a slew of arguments against animism. Most of his theological arguments are fine-grained disputes with Christian theologians, both Catholic and Protestant, who found the vulgarly received notion of nature as an intelligent being not only compatible with religion but necessary to it. Like Boyle, these theologians wished to maintain God's honor, glory, and wisdom. The question for them, therefore, was not whether God created the world and was responsible for natural processes, but *how* he created it and how he carried out his responsibility for natural phenomena.

Boyle's many theological arguments in *Some Considerations Concerning the Usefulness of Natural Philosophy* and *A Free Inquiry*, both cited by Sargent, are denials of animism in favor of mechanistic accounts of God's creation and of his relationship to the world. And none of the considerations Sargent presents support the idea that Boyle had a natural theology that was logically independent of his objections to animism in favor of mechanism, for his arguments proceed by setting mechanism against an objectionable animism.

Boyle finds the "most insidious" use of the term "nature" to be in reference to "a goddess or a kind of semi-deity." And, as Sargent points out, he argues that although it is acceptable linguistic usage to say "that nature does this or that, we ought not to suppose that the effect is produced by a distinct or separate being." We ought not to do so because, in Sargent's words, the "linguistic confusion that gave rise to the idea of nature as a type of purposive agent was a theological threat, because it could result in a failure to appreciate that, by his unlimited power, God created the world and all its processes without the assistance of a 'vice-regent'" (Sargent, *Diffident Naturalist*, 94). Here, in Sargent's own words, Boyle is arguing that God is more powerful according to mechanism (he creates the world alone) than according to animism (he uses a subordinate to do the work). Contrary to Sargent's claim, he does not argue for a theological view of God's power independently of mechanism; rather, the argument is that mechanism supports proper religion because it conduces more to God's power than does animism, which can lead to failure to appreciate this attribute of God.

Boyle's arguments for determinism move in the same way. His commitment to physical determinism is found most clearly in his analogies between the world and an engine and between the world and a clock. As Sargent points out, in opposition to the view that nature interposes herself whenever an emergency arises, Boyle says that the world "is like a rare clock . . . where all things are so skilfully contrived, that the engine being once set a moving, all things proceed, according to the artificer's first design, and the motions . . . do not require, like those of puppets, the peculiar interposing of the artificer, or any intelligent agent, employed by him, but perform their functions upon particular occasions, by virtue of the general and primitive contrivance of the whole engine" (Sargent, *Diffident Naturalist*, 99–100). Again, the very passage Sargent quotes makes clear that Boyle is working out his account of creation as mechanistic and deterministic in direct opposition to animism. Thus, Boyle's theological account of creation was not logically prior to his mechanism (or independent of his objection to animism).

Finally, Sargent mentions Boyle's objection to using the term "nature" for an intelligent being: that it was "no small impediment to the progress of sound philosophy." And, she says, "In place of the 'custom of assigning, as true causes of physical effects, imaginary things or perhaps arbitrary names,' he wished to discover the true locus of causality, which would have to make reference to physical agents" (Sargent, *Diffident Naturalist*, 94). Boyle certainly wanted to replace animistic causal entities with physical causes, but if readers turn the page following her quotation, they find clear evidence that Boyle's theology and his mechanism were developed together, for Boyle argues that animism is theologically and politically pernicious and mechanism is not, so mechanism provides a better support for (proper) religion:

As for religion . . . I hope the doctrine we have proposed may appear fit to do it a threefold service.

I. And in the first place, our doctrine may keep many, that

were wont, or are inclined, to have an excessive veneration for what they call nature, from running, or being seduced, into those extravagant and sacrilegious errors [lately revived].

The first of these is pantheism; these people acknowledge God,

but mean such a God, as they often too little discriminated from matter, and even from the (a) world. . . . [Whereas our God is, firstly, perfect; secondly,] both incorporeal and too excellent to be so united to matter, as to animate it like the heathen's mundane soul; or to become to any body a soul properly so called; and thirdly, uncapable of being divided, and having either human souls or other beings, as it were, torn or carved out . . . so as to be truly parts or portions of his own substance. (b) Whereas, the idolaters and infidels I speak of conceived, under the name of God, a being [which is purely corporeal; here Boyle also mentions Hobbes and "the ancient Aegyptian theologers"]. But secondly, there are others, that allowed a soul of the world, which was a rational and provident being, together with the corporeal part of the universe, especially heaven. . . . but withal, they held, that this being did properly inform this great mass of the universe, and so was, indeed, a mundane soul. And though some of our late infidels (formerly pointed at in this treatise) pretend to be great discoverers of new light in this affair; yet, as far as I am informed of their doctrine, it has much affinity with, and is little or not at all better than, that . . . asserted by the Stoics. (W, V, 250–251)

Thus, Boyle says that "our doctrine" serves religion by keeping credulous folk from pantheism and from belief in a World Soul and thus from the political path of "our late infidels." This is not to say, with Sargent, that a politically innocent theology led to Boyle's mechanism.

The corpuscular philosophy is preferable to animistic natural philosophy because the latter can lead the unwary into heresies and thence to behave like "our late infidels." This is part of the story, but it is only *part* of the story. Boyle also adduced considerable experimental support for the corpuscular philosophy and attacked experimental support for hylozoism.

14
Experimental Support for
the Corpuscular Philosophy

The traditional view has been that experimental data alone led Boyle to pursue a mechanistic research program. (And even updated versions of the view have made no mention of Boyle's deep interest in and debt to Hermeticism or the political dangers facing such an interest, and have not connected these dangers to Boyle's own objection to hylozoic natural philosophy and his explanation of the corpuscular philosophy's advantages. We will deal with such a version in chapter 18.) But was mechanism obviously superior to hylozoism at the time Boyle took it up? In retrospect, we are tempted to say, "of course it was." But we look back across three hundred years of work on problems generated by mechanism. In fact, strict mechanism, as Henry and Clericuzio have argued, was probably never held by anyone but Hobbes and Descartes, if by them. Magnetism and gravity were not satisfactorily explained on a strictly mechanical paradigm; to give an account of gravity, Newton had recourse to a (then) very occult notion, force. And in biology, vitalism lived on into the nineteenth century. In the 1640s and '50s, the strongest support for mechanism was to be found in mechanics (e.g., in the work of Galileo) and in statics. In mid-sixteenth-century England, hylozoic medicine (based in Galen's work) faced a strong challenge from hylozoic iatrochemistry (based in the work of Paracelsians and Helmontians). But in England as in Europe, hylozoic mechanics (based in Aristotelian physics) faced its strongest challenge from mechanism (based in the work of Galileo, Descartes, and Gassendi).

Thus, Boyle argues in *A Free Enquiry* that the hylozoic paradigm does not explain natural phenomena as well as the corpuscular philosophy does: "I observe divers phaenomena, which do not agree with the notion or representation of nature, that I question." Those who hold the hylozoic paradigm, he notes, appeal to a set of examples "taken from bodies destitute of life" (of course, whether the bodies are destitute of life is the question at issue); these include "the ascension of water in sucking pumps, and the sustentation of it in gardeners watering pots [clepsydra]" alleged to be caused "by nature's abhorrence of a vacuum"; heavy bodies falling to the ground in a straight line because "nature directs them the shortest way to the centre of the earth"; bubbles rising through water and flames ascending "because nature directs the bodies to rejoin themselves to their respective elements" (i.e., air and fire move upward toward their natural places: air above the earth and fire above the air). And Boyle attacks the hylozoic explanation of these phenomena by describing experiments whose results he alleges to be inconsistent with it; for example, a ball dropped from a height bounces and does not try to fall directly to the center of the earth as all earthy bodies are supposed to "desire" to do (*W, V*, 192 and 194).

He understood the most devastating of these counter-experiments to be the Torricelli experiment showing that Nature is unable to raise the mercury in a Torricelli tube even a finger's breadth above thirty inches, contrary to the claim of the Aristotelians that Nature would raise a liquid to any height necessary to prevent a vacuum. This anomaly (and related ones) had vexed Aristotelian physicists; and given its central place in the mid-seventeenth-century controversy among competing physics, Franciscus Linus's ingenious solution to it was bound to be contested. Between writing *The Usefulness of Natural Philosophy* and *A Free Enquiry,* Boyle struggled with proponents of rival views over the proper interpretation of the Torricelli experiment, similar experiments with the J tube, and mercury experiments performed using the pneumatical engine. As we saw in chapter 2, he responded to three alternative accounts of the Torricelli experiment, those of Franciscus Linus, Thomas Hobbes, and Henry Moore.

A Defence of the Doctrine Touching the Spring and Weight of the Air . . . Against the Objections of Franciscus Linus not only defends Boyle's corpuscular philosophy, and especially its interpretation of the Torricelli experiment, but also attacks Linus's "vulgar" alternative—one conducive to the animism Boyle had found so objectionable in *The Usefulness* and to which he objected again in *A Free Enquiry.* The arguments in both of these works show us that Boyle understood the ideological importance of his refutation of Linus in *A Defence,* even though he does not mention the connections among Linus's Aristotelian or "vulgar" philosophy, animism, and "our late infidels." At stake in his dispute with Linus, then, were both the proper explanation of physical phenomena and, *inter alia,* proper gender and class relations among the people of England. For our purposes, however, Linus is not the ideal opponent. Though Boyle attacks the animism behind Linus's view that a vacuum is impossible, Linus certainly does not represent the lively animism of a Campanella or Helmont. Both were atomists and both vacuists: Campanella thought it possible to create a vacuum and Helmont posited an interstitial vacuum, empty of body but containing *magnale.* Since Boyle clearly admired Helmont's careful experimentalism, he would have been a much more interesting opponent for our purposes. But Linus is the representative of animism with whom Boyle went toe-to-toe over interpretation of the experimental data and it was in trying to refute Linus that Boyle produced the Gas Law.

Against Linus's attack, Boyle defends his own hypothesis (referred to variously as the "Elaterist" and the "Epicurean" hypothesis) by showing that he can give "a good account" of several experiments Linus claims Boyle cannot account for. Most notably, Linus argues against Boyle's fundamental claim that the weight and spring of the air explain the Torricelli experiments. Linus says, Boyle tells us, that

> if a tube but twenty inches long . . . be not quite filled with quicksilver, as before, but a little space be left betwixt the mercury and the finger on the top of the tube, in which air only may abide; we shall find that the finger below being removed, the finger on the top will not only be drawn downwards, as before, but the quicksilver shall descend also, and that notably, viz. as much as so small a parcel of air can be extended by such a descending weight. So that if, instead of air, water, or any other liquor which is not so easily extended, be put in its place, there will be no descent at all.
>
> Hence, I say, against this [Boyle's] opinion an argument is framed: for if the external air cannot keep up those twenty inches of quicksilver from descending, as we have proved; how shall it keep up twenty-nine inches and an half? Assuredly these can no way be reconciled. (*W,* I, 125)

(See Figure 1.) Throughout these discussions, it is important to remember that Linus did not deny that the air has some weight and some spring; rather, he denied that it has enough weight and spring to account for the experiments under consideration. In the passage quoted here, Linus argues that there is no column of air pressing on the mercury in the tube because one's finger is resting on the top; the only air is that bit at the tip of the tube between one's finger and the surface of the mercury. When one's other finger is removed from the base of the tube, the mercury falls despite the fact that it is less than 29½ inches to begin with—for Boyle had claimed that the external air pressing on the mercury in the dish would hold the mercury 29½ inches up in the tube. Linus cleverly points out that here it cannot even hold up twenty inches of mercury, so we cannot plausibly believe it could hold up more. Instead, Linus's alternative explanation is that the "small parcel" of air at the tip of the tube forms a membrane, or funiculus, and extends as far as it can, holding the mercury up as high as it can. If water were at the tip instead of air, the mercury would not fall at all, since any membrane formed from water is not as easily extended as membranes formed from air.

Boyle's response is that the bit of air at the tip of the tube expands and "thrusts down" the mercury in the tube. But Linus had foreseen this response: "So should the finger be rather thrust from the top of the tube, than thereby fastened to it; because this dilatation must be made as well upwards as downwards" (W, I, 126). To this, of course, Boyle's response is that "the whole weight and pressure of the atmosphere" rests on the finger, pressing it down into the tube and overcoming the pressure of the air in the tip of the tube, even though that air is aided by the pressure of the atmosphere as well, in addition to the weight of the mercury pressing back against it (that is, the atmosphere is pressing upon the dish of mercury, pushing the mercury back up the tube against the air and against the finger).

In Experiment XVII, Boyle describes the void-in-a-void experiment: one end of a glass tube is sealed, the tube is filled with mercury, and then the open end of the tube is covered by a finger. The covered end is set into a basin of mercury and the finger is removed. The height of the mercury drops a bit, leaving a space at the tip of the tube. The basin and the tube are then placed in the receiver of the air-pump, and we observe that the height of the mercury in the tube and in the basin varies as air is pumped from and let into the receiver. Boyle argues that variations in the pressure of the air cause the height of the mercury to vary. Linus answered that as the air is pumped from the receiver, it is "greatly rarified *and extended*"; that is, it forms into a membrane and "vehemently contracts itself," so that it lifts the mercury in the basin, which in turn allows the mercury in the tube to descend, "so that it is no wonder, that the external air afterwards entering, the mercury in the tube ascends," since the air in the receiver is no longer extended, no longer contracts, and no longer lifts the mercury in the basin (*W*, I, 168).

Our analysis of these exchanges reveals that Boyle and Linus are both able to explain the experimental data fairly well. We shall see as we analyze Boyle's *Defence* that his dispute with Linus provides an all-too-apt example of the frustrations that attend attempts to adjudicate between fundamentally different scientific views. Indeed, Boyle himself notes at the outset of his *Defence* that Linus "assumes those very things as principles, that to me seem almost as great inconveniences as I would desire to shew any opinion I dislike, to be liable unto." We are not surprised, then, to find that Boyle uses other criteria in addition to empirical adequacy. For example, unintelligibility turns out to be fundamental among Boyle's attacks upon the funicular hypothesis; he will, he says, show this hypothesis to be "partly precarious, partly unintelligible, partly insufficient, and besides needless" (*W*, I, 124 and 134). By "intelligible" Boyle generally means mechanical. Thus, he criticizes scholastic explanations in terms of "real qualities," "natural powers," and "faculties" because "they do not intelligibly declare what this faculty is and in what manner the operations they ascribe to it are performed by it." As an example of an intelligible explanation of a faculty, such as a magnet's attractive faculty, Boyle refers to his changing the texture of a magnet, thereby showing that "even this wonderful attractive faculty" depends "upon the mechanical structure of the mineral and its relation to other bodies among which it is placed, especially the globe of the earth and its magnetical effluvia" (*W*, V, 247). In his *Excellency and Grounds of the Corpuscular or Mechanical Philosophy*, he identifies intelligible causes with physical causes and gives both a mechanical gloss, saying that mechanical philosophers are satisfied that one part of matter can act upon another only by local motion or by the effects of local motion; so that "if the proposed agent be not intelligible and physical, it can never physically explain the phaenomena; so, if it be intelligible and physical, it will be reducible to matter, and some or other of those only catholick affections of matter already often mentioned [e.g., shape, size, and texture of corpuscles]" (*W*, IV, 73). We will discuss the other criteria Boyle mentions below.

Not all the experiments at issue between Boyle and Linus end in a draw from our point of view. Linus had a tough time with some, and Boyle had a tough time with others. One of Boyle's greatest challenges was to give a corpuscular explanation of the phenomenon of cohering marbles. Two smooth discs of marble (or glass), when pressed together, spontaneously cohere and can be pulled apart only by great force. Since antiquity, thinkers had used this phenomenon to argue over the possibility of a vacuum. Roughly, vacuists argued that since there is no air between cohering discs, when they are pulled apart a vacuum must exist between the discs before the air rushing in from all sides reaches midpoint. Plenists argued that the discs cohere so strongly because of nature's horror of a vacuum and they found various ways to explain the separation without allowing the existence of a vacuum. (For example, Hobbes argued that the discs flex, thereby allowing air to run through the space between them.)

To be accounted successful, physical theories had to explain this phenomenon. Boyle's explanation for why, when the top disc is lifted, the bottom disc does not fall was that, although the upper disc prevents the air from pressing down on the lower disc, nevertheless the air presses up against the bottom of the lower disc and so holds it against the upper one. Accordingly, he expected that when cohering discs were placed in the receiver of an air pump and the receiver evacuated, the lower disc would separate and fall, since it would no longer be sustained by the air pressure. (See Figure 5.) But he reports in *New Experiments* that when he tried the experiment, the lower stone did not fall away, even when a four-ounce weight was added to it. Boyle explained his failure by pointing out how hard it was to get such marbles ground sufficiently smooth, and how difficult it was to find the proper liquid to smear on their irregular surfaces so as to make them smooth enough to cohere. But he transformed failure into success by arguing that the receiver leaked, so it was not really empty

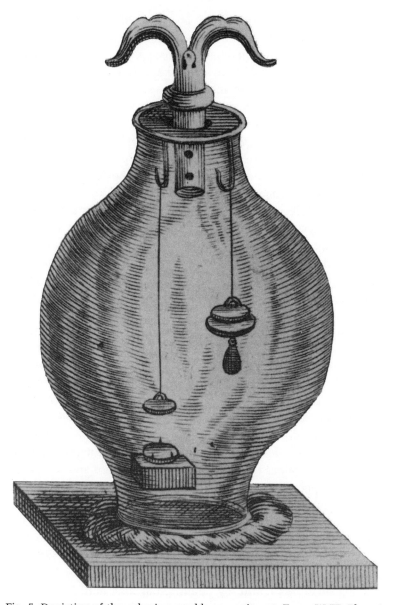

Fig. 5. Depiction of the cohering marbles experiment. From *W*, III, Plate 4.

and enough air remained in it to press the lower stone up against the upper stone. That is, he argued that the experiment demonstrated not only that the air has pressure, but that even rarified air has very strong pressure (*W*, I, 69).

Although Linus agreed that the air has some weight and spring, he did not think them sufficient to hold up the lower disc. The weight and spring of the small amount of air leaking into the receiver would, then, certainly be insufficient. Therefore, he refused to accept Boyle's explanation for the failure of the marbles to separate: "It is certain, that that opinion [i.e., Boyle's hypothesis that the air has great pressure] *is sufficiently refuted by this single experiment*" (*W*, I, 173; emphasis mine). Boyle responded that if Linus had done all the experiments with discs that Boyle had done, "possibly he would have spoken less resolutely." And in further defense of his hypothesis that the air has weight and spring sufficient to cause these experimental phenomena, Boyle referred to Experiment XVII, which he had interpreted according to that hypothesis, and tried to give a fuller picture of how the weight of the air forces the lower disc against the upper one. A column of air whose diameter is the same as the diameter of the disc weighs down, he said, so that it pushes up on the bottom of the lower disc. (Since the column of air begins at the top of the atmosphere, presumably the lines of force by which it exerts pressure on the bottom of the lower disc would take the shape of a J. Boyle never explains how the weight of the air does this. In other texts, he explains that the air's *spring* causes the lower disc to press upward. This explanation is more satisfactory to us, but as we know, he did not give an explanation of the air's spring.)

Linus gave a different explanation for why it is difficult to separate cohering discs and how they eventually come to separate:

Since, when falling, the lower stone would have to disconnect simultaneously from the whole of the upper surface, nor could the neighbouring air [simultaneously] insinuate itself into the whole remaining space, it is necessary that the stone would descend in no other way than by leaving after itself a fine substance, which mercury or water leave behind them when descending this way. Yet since such a substance is separated from the marble with more difficulty than from mercury, or any

other fluid body, it thence follows that it adheres here so tena-
ciously.[1]

Since Nature abhors a vacuum and a vacuum would otherwise be
created when the marbles are forcibly separated, it must be that a
funiculus or "fine substance" is formed from the marbles as they separate and fills the space between them. The discs cohere so strongly
because it is much more difficult for a funiculus to form from marble
and other hard bodies than from liquids.

In the *Defence,* Boyle argues that Linus's funicular explanation is
"precarious." Elaborating, he says,

> although I have tried experiments of this nature with stones of
> several sizes . . . yet I never could find that by their cohesion
> they would sustain a weight [appended to the lower stone]
> greater than that of a pillar of the atmosphere that pressed
> against the lowermost. (*W,* I, 174)

Thus, Boyle thinks that the amount of weight the lower disc would
sustain before separating from the upper disc could be a measure of
the pressure of the air in the receiver, i.e., the more weight it takes to
make the disc drop off, the greater the air pressure. Thus, in this response to Linus, Boyle unfortunately assumes not only that which is
to be proven, viz., that the air has enough pressure to sustain cohering marbles, but also that he has found a way to measure the air's
pressure.

Boyle sandwiches this experiment into his *New Experiments* as number XXXI and devotes very little space to it; without knowing the history of cohesion, one would think Experiment XXXI of little importance. But he was certainly embarrassed by its failure and worked
on it for the next fifteen years. Soon after publishing his responses
to Linus and Hobbes, he used a new air pump with a new receiver
and apparatus (including much heavier weights—up to a pound—
appended to the lower disc) to perform the experiment, and he declared his success in *Continuation of New Experiments,* published in
1669. In this work, he describes adding a one-pound weight to the
lower marble, emptying the receiver of air, and seeing the disc fall
down.[2]

Although Boyle had problems with the cohering marbles, most of the experimental data he produced can be reasonably well explained by his hypothesis as well as by Linus's. The data, as we see, do not clearly favor one or the other fundamental hypothesis. Thus, when Boyle attacks Linus's positive arguments for the funicular hypothesis in Part II of the *Defence,* he points out that they rest upon Linus's belief that a vacuum is impossible:

> according to the examiner, there can be no vacuum; and that
> he makes to be the main reason why nature in the Torricellian
> and our experiments does act after so extraordinary a manner
> as is requisite to the production of his Funiculus. (*W,* I, 135)

For, Linus says, "this comes to pass [i.e., a membrane is produced] that there may be no vacuity, seeing there is nothing else there that can succeed into the place of the descending quicksilver. And hence is confirmed that common axiom used in the schools for so many ages past, that nature doth abhor a vacuum." Boyle's response is to demand a proof of his opponent's fundamental principle: "I see not how our examiner cogently proves, either that there can be none [no vacuum] *in rerum natura,* or that *de facto* there is none produced in these experiments" (*W,* I, 135). Of course, Linus could rightly respond that neither has Boyle cogently proven that a vacuum is possible or that one has ever been produced. The problem is that both sides can appeal to the same experiments for supporting evidence; both produce interpretations of the data that adequately explain them. We see this yet again in Linus's argument that the tip of the Torricelli tube does not contain a vacuum because if it did, it "would appear like a little black pillar because no visual species could proceed either from it, or through it, unto the eye." That is, Linus held an Aristotelian theory of light, according to which, if a vacuum existed, light could not pass through it, because light is a substance providing the medium through which things like the inside of the tip of the tube become visible to us.

At one point, Boyle actually admits that the experiment does not prove either a vacuum or a funiculus: "the experiment will not demonstrate that there is nothing of body in any part of the space deserted by the mercury, so neither will the argument conclude . . . that space . . . to be filled with any true substance." But he gives two atomist theories of light by which to explain why we do not see a black pillar:

light consists of atoms small enough to penetrate the glass walls of the tube, or light is "a propagation of the impulse of lucid bodies" through the few remaining particles of matter in the tip of the tube (*W*, I, 135–136). We might note that on neither theory is there an absolute vacuum in the tip of the tube; to this extent, then, Boyle fails to refute the Aristotelian hypothesis that light cannot travel through a vacuum. Nevertheless, Boyle concludes that his arguments show Linus's funicular hypothesis to be "precarious" because "his arguments do not sufficiently prove" that there is no vacuum in the tip of the tube and the funicular hypothesis depends on having done so. But we might, again, say the same of Boyle's arguments for a vacuum.

As we have seen, Boyle's objection to nature's abhorrence of a vacuum is a corollary of his objection to animism. This appears clearly in his attack on Linus for failing to explain the "inward spring" of the membrane (i.e., how the membrane can pull mercury up toward itself):

> It seems also very difficult to conceive, how this extenuated substance should acquire so strong a spring inward, as the examiner all along his books ascribes to it. Nor will it serve his turn to require of us in exchange an explication of the air's spring outward, since he acknowledges, as well as we, that it has such a spring. [Hobbes had charged Boyle with hylozoism because he failed to explain the air's outward spring, and without an explanation, the notion of spring is inconceivable.[3]] . . . Nor can it well be pretended, that this self-contraction is done *ob fugam vacui,* since though it should not be made, a vacuum would not ensue. And if it be said, that it is made, that the praeternaturally stretched body might restore itself to its natural dimensions; I answer, that I am not very forward to allow acting for ends to bodies inanimate, and consequently devoid of knowledge; and therefore should gladly see some unquestionable examples produced of operations of that nature. (*W*, I, 143)

In the remarks quoted here, Boyle claims that a membrane is not necessary even on a hylozoist account because on that account a vacuum is impossible and would not occur whether or not the funiculus was produced by the mercury. But this claim merely marks the difficulties that arise when scientists argue across fundamentally different theories, for the claim amounts to a denial of the fundamental hylozoic principle that Nature always takes care that a vacuum does not occur—which means that some phenomenon will always occur instead of a vacuum. This principle allows scientists to look for such phenomena and expect to find them; and Linus argues that in this case, the membrane is the phenomenon that prevents a vacuum.

But by now we know that, though Boyle understates it, his real objection is to hylozoism itself: "I am not very forward to allow acting for ends to bodies inanimate . . . and therefore should gladly see some unquestionable examples produced of operations of that nature," he says. Given the strength of that objection and of his commitment to mechanism, one wonders what an "unquestionable example" of hylozoism could possibly be for Boyle. From a mechanistic point of view, hylozoic explanations will always be, as Boyle remarked, "unintelligible" and "very difficult to conceive."

This is nowhere clearer than in Boyle's attack upon the Aristotelian concept of rarefaction that Linus depends on when he describes the production of a funiculus. This concept was notoriously problematic even before the seventeenth century, even to Aristotelians, but we can, I think, make rough sense of it. Aristotelians imagined the rarefaction and condensation of bodies to be changes in the property of density and conceived the property of density in some of the ways that we conceive the property of shape; for them, the density of bodies was changeable in the way that shape is changeable. Thus, a liquid like water or mercury could change its density by taking up more or less space as easily as it changed its shape, and in neither case did its weight, or any of its other properties, change. In neither case was empty space of any kind necessarily created inside the body, nor did any other substance enter the body. Once again we need to be careful here not to assume, because the early modern corpuscular account of matter is so similar to our own in its explanation of density, that it was absurd of Linus and other Aristotelians to hold on to the Aristotelian notion of density. As Boyle notes in another connection, an explanation might be forthcoming at any time, just as he himself might be unable to explain the air's spring, but someone else might in the future. Just so, Aristotelians could hope for a future explanation of their concept of density.

Boyle argues that the Aristotelian concept of rarefaction is unintel-

ligible to him, "as well as to many other considering persons." But we
see that his objections to it all work by demanding that this Aristotelian notion be explicable in terms of other theories, either Cartesian
plenism or Epicurean vacuism. Here we will examine only two of
Boyle's objections. Linus and other Aristotelians argued that a small
volume of air can expand to fill a space two hundred thousand times
larger without admitting any vacuities or any other substances into
it. Boyle's response is that this account

> will questionless appear very absurd to the Cartesians and
> those other philosophers, who take extension to be but notion-
> ally different from body, and consequently impossible to be ac-
> quired or lost without the addition or detraction of matter; and
> will, I doubt not, appear strange to those other readers, who
> consider how generally naturalists have looked upon extension
> as inseparable, and as immediately flowing from matter; and
> upon bodies, as having necessary relation to a commensurate
> space. (*W,* I, 145)

The Cartesian notion of space is logically (biconditionally) dependent
upon Descartes's notion of extension, which can be explicated as follows; Descartes identified matter with extension. That identification
means that all material bodies are extended, and every space either
is a body or is a possible place for a body (though at the moment it is
full of some other body). It follows that there is no empty space, of
course, but it also follows that any space has a body in it commensu-
rate with that space. Consequently, if one space A is larger than another space B, it is impossible for the same body to fill A at one time
and B at a later time. Hence, Linus's Aristotelian notion of rarefaction
and condensation is absurd. (And so was Gassendi's notion and probably Boyle's notion—as well as our contemporary notion—inasmuch
as they all depend upon the possibility of empty space.) In the same
way, Boyle's attempt to reduce Linus's notion of rarefaction to the absurd also depends upon a Cartesian notion of space. Linus's concept
of rarefaction is absurd, Boyle says, because according to Linus's notion it is possible for the world to "be made I know not how many
thousand times bigger than it is, without either admitting any thing
of vacuity betwixt its parts, or being increased with the addition of
one atom of new matter" (*W,* I, 145). But one man's *reductio* is another

man's *modus ponens*. Aristotelians simply admit that it is possible but very unlikely. However, from Boyle's corpuscular point of view, any such rarefaction supposes a miracle: God (or, though Boyle eschews this possibility, some other divine or semi-divine being) must create new matter to fill the larger space. Thus he remarks,

> For though God can both create and annihilate, yet nature can do neither: and in the judgment of true philosophers, I suppose our hypothesis would need no other advantage, to make it be preferred before our adversaries, than that in ours things are explicated by the ordinary course of nature, whereas in the other recourse must be had to miracles. (*W*, I, 149)

It is tempting to say that Boyle begs the question against Linus, but the same could be said of Linus; each attacks the other's notion of rarefaction and defends his own from within his own theory. To Boyle's credit, he himself recognizes that the dispute over rarefaction ends in a draw:

> whereas the examiner's argument on this occasion is, that his way of rarefaction must be admitted, because neither of the other two can be well made out; his adversaries may with the same reason argue, that one of theirs is to be allowed, since his is incumbered with such manifest difficulties. And they may enforce what they say by representing, that the inconveniences, that attend his hypothesis about rarefaction, are insuperable, arising from the unintelligible nature of the thing itself; whereas those, to which the other ways are obnoxious, may seem to spring but from men's not having yet discovered, what kind of figures and motions of the small particles may best qualify them to make the body, that consists of them, capable of a competent expansion. (*W*, I, 146–147)

Boyle is referring here to Linus's charge that neither Epicureans (vacuist atomists) nor Cartesians (plenist atomists) can explain rarefaction phenomena as well as he can. Boyle undertakes to explain them and then argues that, although both theories explain the phenomena, Linus's hypothesis is unintelligible, so Linus will never overcome the difficulties facing the Aristotelian account of rarefaction, whereas the difficulties facing Boyle's own account will be overcome in time.

At such an impasse, one hopes for a crucial experiment to decide between two theories having such different accounts of the "rarefaction and condensation" of the air; whether it has "outward spring" or only an ability to "condense." The *experimentum crucis,* he tells his readers, is Pascal's Puy-de-Dôme experiment wherein the Torricelli apparatus was carried up three thousand feet and the mercury in it was reported to have dropped as the apparatus ascended. As we saw in chapter 2, Linus argued that Pascal's results indicated changes in temperature, not changes in the pressure of the air. And Boyle took this counter-suggestion seriously, for in his attempts to replicate Pascal's results by experiments at Westminster Abbey, he tried to answer it in his outdoor experiments and to control for temperature in his indoor experiments. This control was the more necessary since Boyle made use of "a kind of weather-glass" and used water instead of mercury so that "small changes in the weight of resistance of the atmosphere in opposition of the included air might be the more discernable" (*W,* I, 152). A weather-glass is set up so that the water to be measured is in a slender pipe open at the top to the air. According to Boyle's hypothesis, then, the level of water in the pipe should *rise* as the pressure decreases. The problem with the apparatus, as Boyle himself recognized, was that, according to his own theory, it measured both temperature and air pressure. For this reason, he says in another essay, it is not a reliable instrument for measuring the temperature:

> And as to the invention of weather-glasses . . . these open thermometers are not to be safely relied on, since in them the liquor is made to rise and fall, not only, as men have hitherto supposed, by the cold and heat of the ambient air, but (as I have shewn by divers new experiments) according to the varying gravity of the atmosphere; which variation has not only a sensible, but a very considerable influence upon the weather-glass. (*W,* IV, 61)

Though these remarks were written a few years later (in 1665), we can assume Boyle was aware that trying to measure air pressure alone using a weather-glass was problematic. Nevertheless, the experiment was first performed by measuring the level of the water in the apparatus on the lead roof (or "leads") of the Abbey and on the ground, and it was performed once during the day, once during the evening, and again on a later, windy day. Boyle admits that when the wind blew "strongly upon the leads" the results were not as "regular" as before, but he maintains that temperature was not a factor, for

> the water always manifestly fell lower at the foot of the wall than it was at the top: which I see no cause to ascribe barely to the differing temperature of the air above and below, as to heat and cold, since according to the general estimate, the more elevated region of the air is, *ceteris paribus*, colder than that below; which would rather check the greater expansion of the included air at the top of the leads, than to promote it. (*W*, I, 153)

It is likely that other things were not equal and possible that the sun heated the leads on the roof of the Abbey, making it warmer above than on the ground below. The apparatus would have warmed up as it sat on the leads, and cooled again on the ground, for Boyle tells us that he "suffered the glass to rest a pretty while upon the lead." Thus, the results he describes are consistent with Linus's counter-suggestion:

> having suffered the glass to rest a pretty while upon the lead, that the air . . . within might be reduced to the same state, both as to coldness and as to pressure . . . [and] having marked the station of the water F, we gently let down the vessel by a long string to the foot of the wall, where one attended to receive it; who having suffered it to rest upon the ground, cried to us, that it was subsided about an inch below the mark F we had put. Whereupon having ordered him to put a mark at his second station of it E, we drew up the vessel again; and suffering it to rest a while, we observed the water to be re-ascended to or near the first mark F, which was indeed about an inch above E, the other. (*W*, I, 153)

In other words, the experimental results here confirm both the mechanical hypothesis and Linus's counter-hypothesis, for the water level rose at the higher altitude and dropped at the lower altitude. The higher water level might be due to higher temperature or to lower pressure. (Modern barometers would, of course, find no difference in air pressure between the roof of Westminster Abbey and the ground.)

In his indoor experiment, however, Boyle planned to control for temperature by using a second weather-glass "so contrived, that the weight or pressure of the atmosphere should make no change in it" and which would presumably, therefore, change only in response to changes in temperature. It is difficult to make out exactly why this important experiment was not made. For one thing, the glass apparatus broke, but Boyle's remarks indicate that there were other problems: "there happened in our trials a circumstance or two that seemed not so devoid of difficulties, but that we think it may require further examination," so he says he plans to describe this experiment "as how it succeeds with quicksilver instead of water," in an appendix to *New Experiments*. Since we are without this description, we do not know whether the "difficulties" were technical ones or whether the results were disappointing. But "it shall suffice us, in the mean time," he says, "that the trials already mentioned" support his "doctrine" (*W*, I, 153–154).

To shore up his hypothesis, Boyle argues against Linus that if there were a lower temperature at the top of the mountain than at the bottom, then according to Linus's funicular hypothesis the mercury should stand higher at the top than at the bottom because the membrane would contract in the cold and draw the mercury up (*W*, I, 154). This is a good criticism, the sort of serious criticism that a scientist might use to further develop his hypothesis. In response, Linus might have begun to account for the operations of funiculi under different conditions; generally, the data indicate that some conditions hinder the action of the funiculus and some aid it. He might have hypothesized that cold is among those that hinder. Linus had established that an unhindered funiculus is fully extended by the weight of 29½ inches of mercury at the base of the mountain. He could plausibly have maintained that an unhindered funiculus is fully extended by the weight of 29½ inches of mercury at the top of the mountain as well, but the drop in temperature at such a height (which might otherwise cause the funiculus to contract) causes the mercury in the tube to contract, which, added to its weight, hinders the operation of the funiculus and causes it to rarefy or extend yet more, allowing the level of mercury to drop. And once the thermometer was invented, it should have been possible to quantify precisely the rarefactions and contractions of the funiculus caused by changes in temperature, just as it should have been—and was—possible to quantify precisely the changes in the volume of mercury caused by changes in temperature. We will return to this development of Linus's hypothesis below.

15

Boyle's Law of Gases

The J-tube experiments, in the course of which Boyle produced Boyle's Law, were performed to show that Linus's funicular hypothesis was "needless." Linus admitted that the air has some weight and spring, but not that it has enough to act as a counterweight to 29 inches of mercury. Boyle performed a series of experiments to show that the air has enough weight and spring to counterpose far more than 29 inches of mercury. These experiments were to have the added benefit of showing Linus's funicular hypothesis to be, not just unnecessary, but also inadequate to the data. As we have seen, Linus explained the results of the Torricelli experiment by arguing that the mercury stands at 29½ inches because a funiculus forms as the mercury in the tube drops, connecting the surface of the mercury in the tube to the sealed top of the tube, and the funiculus is unable to hold the heavy mercury any higher than 29½ inches. Against that claim and in support of the hypothesis that the air has great weight and spring, Boyle filled a J tube with enough mercury that the height of the mercury in each arm of the tube was the same; he then measured the length of the short column of air in the tip of the shorter arm. (See Figure 3.) Thereafter, he (or his assistant) poured in enough mercury that the height of the air in the tip of the short arm was "reduced to take up but half the space it possessed (I say, possessed, not filled) before" and the mercury in the longer arm stood about 29 inches higher than that in the shorter arm.

We should note that Boyle does not claim that the results of this operation refuted Linus's funicular hypothesis; instead, he remarks only that the results of his experiment—one volume of air counterbalances a 29-inch cylinder of mercury and half that volume counterbalances twice as much mercury—"agree with and confirm" and "agree rarely-well" with the hypothesis that "the greater the weight is that leans upon the air, the more forcible is its endeavour of dilatation, and consequently its power of resistance." He recognizes that, to refute Linus, he must do more than this. Therefore, he has still more mercury poured into the J tube, until the height of the mercury in the long arm stands at about 100 inches and the volume of air in the tip of the shorter arm is very small. This operation allows him to take the series of measurements that roughly approximated the first half of Boyle's Law. (See Figure 4.) At this point he says, "our adversary may plainly see, that the spring of the air, which he makes so light of, may not only be able to resist the weight of 29 inches, but in some cases of above a hundred inches of quicksilver, and that without the assistance of his Funiculus, which in our present case has nothing to do" (W, I, 159). In other words, this experiment proves, he thinks, that the enclosed air has very great spring; indeed, enough to hold up all that heavy mercury.

But does this experiment show, that is, can any observer plainly see, that the enclosed air has very great spring? Or does it show, so that any observer can plainly see, that the enclosed air is condensed? Linus would have had no problem with the condensation of the air; many phenomena reveal that air, liquids, and some bodies condense under certain conditions, such as when heavy liquids or bodies rest on them. Linus and every seventeenth-century schoolboy had blown air into a carp's bladder and pressed on the bladder; that air condensed until at some point the bladder burst. In Boyle's experiment, heavy mercury is resting on the air and compressing it, and at some point, the tip of the tube will burst.

But the most important problem with this attempt to refute the funicular hypothesis is that a funiculus occurs only to prevent the occurrence of a vacuum. Boyle states that the results of this experiment are explained by the spring of the air and that the funiculus "has nothing to do," and presumably he means that the funicular hypothesis is unnecessary for an explanation and would play no part in the proper explanation. But Linus would have responded that, to begin with, the J-tube experiment is not the same as the Torricelli experiment; in the Torricelli experiment, the funiculus is formed by the surface of the mercury as it falls when the Torricelli tube is inverted. It is formed, probably by mercury vapor, in the space left by the falling mercury; and the mercury only gives it off because otherwise a vacuum would be created. This funiculus then pulls up the weight of

29 inches of mercury. But in the J-tube experiment, the space at the tip of the tube demonstrably contains air; Boyle says so and no one denies it. The space was created by rocking the tube in such a way as to insure that there was air in the tip. Therefore, there is no way or reason for a funiculus to have formed. As Boyle himself says, in the present case, the funiculus has nothing to do.

16
The Production of an Alternative Law

In this chapter, we will see that the experimental data support the production of another scientific law, one more amenable to hylozoism and so to the radical politics that rested upon hylozoism.

Because Linus had said that the funiculus is unable to draw mercury up beyond 29 or 30 inches, Boyle presented a counter-experiment in which mercury is drawn up more than 100 inches. Filling the J tube with mercury, Boyle says,

> we took care, when the mercurial cylinder in the longer leg
> of the pipe was about an hundred inches high, to cause one to
> suck at the open orifice; whereupon (as we expected) the mer-
> cury in the tube did notably ascend. Which considerable
> phaenomenon cannot be ascribed to our examiner's Funiculus,
> since by his own confession that cannot pull up the mercury, if
> the mercurial cylinder be above 29 or 30 inches of mercury.
> And therefore we shall render this reason of it, that the pres-
> sure of the incumbent air being in part taken off by its expand-
> ing itself into the sucker's dilated chest; the imprisoned air was
> thereby enabled to dilate itself manifestly, and repel the mer-
> cury, that comprest it, till there was an equality of force be-
> twixt the strong spring of that comprest air on the one part,
> and the tall mercurial cylinder, together with the contiguous di-
> lated air, on the other part. (*W, I*, 159)

The rising mercury, Boyle says, cannot be explained by the funicular hypothesis because Linus had said that the funiculus is unable to draw mercury up beyond 29 or 30 inches. On the other hand, he is happy to give us the mechanical reason for it, viz., when his assistant sucked air from the top of the tube, he took pressure off both the mercury and the air beneath it so that the air beneath expanded (dilated or rarefied).

But raising a 100-inch column of mercury to a height of 100 + n inches by suction certainly does not refute the funicular hypothesis. More than likely, Linus would simply have stayed with the traditional Aristotelian explanation of suction and argued that as the assistant covered the top of the tube with his mouth and drew the air from it into his lungs, the mercury followed the air upward by attraction, consenting to overcome its natural inclination downward for the good of the whole, i.e., to prevent a vacuum.

But let us suppose Linus (and ourselves) to be persuaded that the conditions in this case were relevantly similar to the conditions produced by the Torricelli experiment; he might then explain that when Boyle's assistant sucked on the tube, a funiculus formed connecting the surface of the mercury to the surface of the assistant's lungs. The funiculus contracted and, aided by the suction from the assistant's lungs, drew the mercury up. Here Linus has an opportunity to develop or elaborate the funicular hypothesis. An unhindered and unaided funiculus can raise a column of mercury to 29½ inches, but under certain auxiliary conditions, it can raise the column higher. In the case under consideration, the funiculus is aided by suction. We have found that the strength of a funiculus is quantifiable (for example, by measuring the height of the column of liquid it holds up) and temperature, suction, and other factors affecting the operation of funiculi are also quantifiable; thus, it will be possible to produce mathematically expressed general funicular laws.

Suppose, for instance, that Boyle had filled the J tube in the same way that one filled and inverted a Torricelli tube (as far as we know, he never did). Linus might then agree that a funiculus had been produced at the tip of the short leg. According to his hypothesis, the funiculus contracts until it is holding up the equivalent of a 29½ inch column of mercury. Why, then, does the little space at the tip of the J tube shrink as more mercury is poured into the tube? (Or when more air is pumped into a receiver containing the Torricelli apparatus?) We may rephrase Linus's hypothesis as follows: an unaided funiculus can lift the weight of 29½ inches of mercury plus the weight of one atmosphere (whatever that weight might be; Linus agrees that the air has some weight, but not enough to account for Boyle's experimental results). And Boyle's experiments with the J tube show that, if the funiculus alone lifts to height H the weight of one atmosphere A plus

29½ inches of mercury M, then when aided by the weight of more mercury or of more air pushing down on the column of mercury, the funiculus can lift the mercury even higher. Linus would make the following table of observations (very similar to Boyle's):

A table of the contraction of the funiculus.

A	B	C	D	E
0	0	H	1At	1M (weight of 29½ inches of mercury)
1M	0	H/2	1At	2M
3M	0	H/4	1At	4M
7M	0	H/8	1At	8M
etc.				
0	1At	H/2	2At	1M
0	3At	H/4	4At	1M
0	7At	H/8	8At	1M
etc.				

A. The weight of the mercury aiding the funiculus.
B. The weight of the air aiding the funiculus.
C. The number of equal spaces in the shorter leg that the funiculus lifted the mercury.
D. The weight of the air weighing down on the mercury in the longer leg.
E. The weight of all the mercury above the funiculus.

And he would be known as the man who discovered an early formulation of *Linus's Law: if air is treated as a liquid, a funiculus lifts to a height proportional to the weight of the liquids it lifts.* The mathematical expression of the relation between the height of the funiculus and the weight it lifts would be $k = hw$. And once the thermometer had been perfected, the mathematical expression would be $k = hwt$.

Of course, what we have done here is modify Linus's theory to cover the data available at the time; theories can be modified by adding auxiliary hypotheses, modifying them in turn, and so on, always maintaining a reasonable degree of empirical adequacy. Whether these modifications are ad hoc depends upon whether the modifications can be generalized to cover other data and used to predict future data, unknown when the modifications were made. We may reasonably suppose that these Linean modifications are generalizable, since they mirror half of Boyle's Law (i.e., the behavior of gases at or above atmospheric pressure, whereby as the pressure increases, the volume decreases). It is, then, in no way clear that Boyle in fact refuted Linus's hypothesis; both hypotheses cover the available data fairly well, and the funicular hypothesis enjoyed the advantage of belonging to the broader Aristotelian world-view, which in the 1650s and 1660s, for all its anomalies, had more explanatory power than the nascent mechanical philosophy.

It might be tempting to conclude with Reilly that because "we hear of no further writings on his part about the subject," Linus accepted Boyle's *Defence* as a refutation of his criticisms. I believe, on the contrary, that examination of the disagreement between them reveals its depth and the firmness with which each of the disputants held to his first principles. If Linus had responded to the *Defence* and Boyle had followed in turn, they would probably have repeated their major arguments—which were well enough set out the first time around. Certainly Linus did not give up his version of Aristotelian physics, for he later defended traditional optics against the young Isaac Newton and maintained a dispute with him cut off only by his death in 1675 at the age of eighty. Moreover, the funicular hypothesis was taken up by others.[1]

17
Methodological Considerations

In this chapter, we examine a number of the considerations advanced by Boyle to support the claim that his corpuscular hypothesis was superior to Linus's funicular hypothesis.

Summarizing his objections to Linus's hypothesis, Boyle says that it is

> partly precarious, partly unintelligible, partly insufficient, and besides needless: though it will not be so convenient to prove each of these apart, because divers of my objections tend to prove the doctrine . . . obnoxious to more than one of the imputed imperfections." (*W,* I, 134, and see I, 177)

Three of these objections reflect a list Boyle once made, setting out criteria for both good and excellent hypotheses:

The Requisites of a *Good Hypothesis* are
1. That it be Intelligible.
2. That it Contain Nothing Impossible or manifestly False.
3. That it Suppose not any thing that is either Unintelligible, Impossible or Absurd.
4. That it be Consistent with it self.
5. That it [be] fit & Sufficient to Explicate the *Phenomena,* especially the Chief.
6. That it be at lest Consistent with the rest of the *Phenomena* it particularly relates to, & do not Contradict any other known Phenomena of Nature, or manifest Physical Truth.

The Qualities & Conditions of an *Excellent Hypothesis* are
1. That it be not Precarious, but have sufficient Grounds in the nature of the Thing itself, or at least be well recommended by some Auxiliary Proofs.
2. That it be the Simplest of all the Good ones we are able to

frame, at lest Containing nothing that is Superfluous or Imperti-
nent.
3. That it be the only Hypothesis that Can explicate the *Phenomena*, or at lest that does explicate them so well.
4. That it enable a skilfull Naturalist to foretell future Phenomena, by their Congruity or Incongruity to it; and especially the
Events of such Expts as are aptly devised to Examine it; as
Things that ought or ought not to be Consequent to it.[1]

A "precarious" hypothesis, then, is without sufficient grounds or
proof, either direct proof, "in the nature of the thing itself," or indirect, through auxiliary hypotheses. Linus's hypothesis is said to be
without enough proof "in the nature of the thing itself." And by "sufficient," Boyle means that a hypothesis explicates the phenomena; we
will take this to mean at least that it accounts for the data. Thus, he
says, Linus's hypothesis does not sufficiently account for the data.

Turning to the charge that Linus's hypothesis is unintelligible, we
have seen that, for Boyle, the requirement that good hypotheses be
"intelligible" means that they should be mechanical. The problem
here is that such a requirement begs the question against Linus and
against any hylozoic hypotheses, for when Boyle says that Linus's hypothesis is unintelligible, he rules out Linus's fundamental explanatory framework. Linus's non-mechanistic assumption that Nature
exists and that she abhors a vacuum, and his funicular hypothesis—
that liquids give off a vapor that thickens into a membrane with
measurable effects on the behavior of those liquids under certain
conditions—might be false hypotheses, but they are not inconceivable. A World Spirit or Nature having consciousness and feelings is no
less intelligible than God's having them. Both were supposed analogous to human consciousness and feelings. And explanations in terms
of a World Spirit or Nature as God's subordinate viceregent directing the activities of, for example, mercury are certainly no less intelligible than God's directing at creation that the universe follow certain
regularities—a claim Boyle never found "unintelligible." Such explanations might be false, but having accepted a Spirit himself, Boyle is
not entitled to reject the World Spirit as unintelligible.

Turning to his conclusion that Linus's hypothesis is insufficient and needless, we find that these failings are interdependent. If Linus's hypothesis is not adequate to account for the data (i.e., is insufficient), then, assuming that Boyle's own hypothesis sufficiently accounts for the data, of course the funiculus is unnecessary (it is needless). But we have seen that both hypotheses possess reasonable empirical adequacy; thus, each author can claim that his own hypothesis is "sufficient" and his opponent's hypothesis is "needless." Boyle's charge that Linus's hypothesis is precarious, that it does not have sufficient grounds in the nature of the thing itself, has a similar problem. Given the animistic assumption that Nature exists and can have the feeling of abhorrence, Linus's hypothesis can be said to be grounded in the nature of the thing itself since it covers the data reasonably well, explaining one of the ways Nature undertakes to avoid a vacuum. On the other hand, assuming that animism is false and mechanism true, Boyle's hypothesis can be said to be grounded in the nature of the thing itself since it also covers the data reasonably well.

We note that Boyle did not claim that his hypothesis was simpler than Linus's. He says Linus's hypothesis is "needless," which is an indirect reference to simplicity; thus, Nature would be an extra explanatory entity, an unnecessary one if Boyle already had the correct explanation of the phenomena. But Boyle did not observe simplicity when he made God an explanation of the origin of motion rather than adopting the Epicureans' simpler view that atoms have always existed and been in motion. Thus, Boyle is caught between a theological rock and hard place. A direct appeal to simplicity as a determining criterion is one way to rule out a World Spirit and animism, but it immediately raises the difficulty that God is unnecessary in the Epicurean hypothesis of matter in motion since the beginning of eternity.

Boyle could claim his view to be simpler than Linus's in that God is only brought in at creation, unlike Nature, who works continually. But Boyle already had a problem that he never satisfactorily resolved, viz., if God does not intervene in the world, there are no miracles and the scriptures are wrong. Moreover, the threat of determinism looms: miracles could be predetermined, i.e., miracles only from our point of view, and the free will of humans is threatened. If, on the other hand, God does intervene, then miracles happen and the scriptures are literally true, but Boyle's ontology is not significantly simpler than Linus's. Boyle sidesteps these issues by arguing only that Linus's hypothesis is "needless."

Although Boyle appears to be satisfied that he has defeated Linus's hypothesis and shown the superiority of his mechanistic one, he nevertheless concludes the *Defence* by remarking,

> But this notwithstanding, I am . . . content, my adversary
> should be thought to have said for his principles as much as
> the subject will bear; nor would I have it made his disparage-
> ment, that I have declared that his whole book has not made
> me depart from any of my opinions or explications, since his
> hypothesis and mine being inconsistent, it may be looked upon
> as a sign rather that each of us have, than that either of us
> have not, reasoned closely to his own principles, that the
> things we infer from our contrary suppositions do so generally
> disagree. (*W,* I, 178)

Given that hylozoism and mechanism are contrary fundamental principles, inferences drawn from each will be inconsistent.

Laudan finds in such remarks evidence for Boyle's holding an early version of the underdetermination thesis: "Boyle does not believe that theories will arise ready-made from the data, or that the data will uniquely determine any single theory."[2] And, indeed, Boyle is quite clear on this issue, for he also says that

> as confidently as many atomists, and other naturalists, presume
> to know the true and genuine causes of the things they attempt
> to explicate; yet very often the utmost they can attain to, in
> their explications, is, that the explicated phaenomena may be
> produced after such a manner, as they deliver, but not that
> they really are so. For as an artificer can set all the wheels of a
> clock a going, as well with springs as with weights; and may
> with violence discharge a bullet out of the barrel of a gun, not
> only by means of gunpowder, but of compressed air, and even
> of a spring: so the same effects may be produced by divers
> causes different from one another; and it will oftentimes be
> very difficult, if not impossible, for our dim reasons to discern
> surely, which of these several ways, whereby it is possible for

nature to produce the same phaenomena, she has really made use of to exhibit them. . . . it is a very easy mistake for men to conclude, that because an effect may be produced by such determinate causes, it must be so, or actually is so. And as confident as those we speak of [the ancient atomists] use to be, of knowing the true and adequate causes of things, yet Epicurus himself, as appears by ancient testimony, and by his own writings, was more modest, not only contenting himself, on many occasions, to propose several possible ways, whereby a phaenomenon may be accounted for, but sometimes seeming to dislike the so pitching upon any one explication, as to exclude and reject all others. (*W*, II, 45; see also *W*, I, 307)

These important statements reveal Boyle's sophisticated position on the question of whether a set of data determines its unique interpretation. Many causes could be assigned for some things, he tells us. The actual explanation of phenomena under investigation might always differ from the explanation we have hit upon; it is always possible for two or more such explanations to account for the phenomena adequately.

Laudan has distinguished two senses of underdetermination: (a) strict underdetermination, whereby two or more hypotheses are (perpetually) evidentially equivalent in that no conceivable evidence could ever discriminate between them, and (b) what we might call historical underdetermination, whereby current rules and existing evidence do not allow us to decide between two hypotheses; although we can conceive of evidence that would differentially support one hypothesis, we do not now have it.[3] Thus, the hypotheses presented by Boyle and Linus around 1661 were historically underdetermined; that is, the experimental and observational evidence presented at the time did not determine whose hypothesis was correct. (I do not claim that their hypotheses were strictly underdetermined, a notion I have dealt with elsewhere.[4])

As we know, *New Experiments Physico-Mechanical Touching the Spring of the Air* and the *Defence* were written to present a set of experiments explained by and supporting the mechanical philosophy and to defend mechanism. Boyle had already told us (in *The Usefulness of Natural Philosophy*) that he objected to hylozoic explanations, and his

air-pump experiments, among others, show that mechanistic explanations "agree with and confirm" and "agree rarely-well" with the
data. But without the criterion of "intelligibility," which amounts to
a prior commitment to mechanism, Boyle's arguments in the *Defence*
are much weaker. His appeal to "intelligibility" and "need" shows us
just how important such criteria are when the data do not uniquely
determine one hypothesis. We will return to this point below.

18

"The Data Alone Proved
Boyle's Hypothesis"

In this chapter, we will examine important recent scholarship arguing that although science is social in many important senses, Boyle chose to pursue the mechanical philosophy and produced the law named for him using only a method based on experimental and observational evidence, and that he was not influenced by social and political considerations such as those advanced in this book.

From the point of view of contemporary science, it is tempting to ask whether Boyle adopted the mechanical philosophy and then recognized its ideological consequences or adopted the theory because of its ideological consequences. These questions are not very helpful, since they presuppose only two possibilities: the scientist adopts a hypothesis either independently of ideological considerations (a good, unbiased scientist) or because of ideological considerations (a bad, biased scientist). And this binary set of questions and assumptions is reflected in the traditional philosophical distinction between the context of justification and the context of discovery. Against the claim that contextual values sometimes influence good science, traditionalists have argued that even though the inception and reception of scientific theories are vulnerable to contextual values, the technical content of those theories, e.g., as expressed in the classical content of Boyle's Law of Gases, k = pvt, is not influenced by considerations of gender or any other politics. In philosophy, this defense rests upon the distinction between the context of discovery and the context of justification, a distinction first put forward early in this century by Hans Reichenbach.[1] This distinction allows defenders of science to admit that scientific problems and solution outlines may be determined by, for example, the social values of granting agencies, as the problem of cancer is currently seen as a genetic problem to be solved by technologies of molecular biology and not as an environmental problem to be solved by technologies insuring that our air, water, and food are not polluted with carcinogens. This is merely the context of discovery. Regardless of how the theories of molecular cell biology came to be discovered and regardless of their later use or misuse, the laboratory work proving them—the context of justification—is value-neutral because it is governed by methods designed to weed out personal and political bias.

Traditional philosophy of science would "save" Boyle as a "good, unbiased scientist" in the face of the connection between his ideological concerns and his scientific ones by relegating the influence of his ideological concerns to the context of the discovery of his hypotheses and arguing that no contextual values were operating in the context of justification. If we point out that since contextual values influenced Boyle's pursuit of a mechanistic research program, those values ultimately influenced his interpretation of the experiments leading to Boyle's Law, traditionalists respond as follows: Even though the historical facts lead to this conclusion, we can rationally reconstruct history to fit a proper account of how scientific work should proceed. (These accounts used to be called logics of science.) So, regardless of whether scientists actually follow proper norms, their work can be rationally reconstructed to fit those norms. This reconstruction or internal account describes the context of justification, and any irrational, biased behavior can be relegated to the context of discovery (to be described by the historian or sociologist of science).

We see that the distinction between the context of discovery and the context of justification is not useful for writing empirically adequate history of science; but its purpose was to maintain the myth that all good scientific work is "rational," where this means free of contextual values. Fortunately, we do not need this distinction to understand the rationality of science. A more historically adequate account of Boyle's pursuit of a mechanistic research program and his production of Boyle's Law allows us to see that the binary questions and assumptions above oversimplify and seriously distort the history of Boyle's work. Seventeenth-century virtuosi understood ideological consequences to be among the criteria not only for good theories but, indeed, for true theories. The thesis adopted here is that both ideological consequences and empirical adequacy were among the considerations that led him to pursue the corpuscular philosophy and to develop it in the way that he did.

Traditional Boyle scholars deny this and argue that, on the contrary, Boyle pursued the corpuscular philosophy simply because his experimental data proved it. That is, as Boyle might have said (but didn't), the corpuscular philosophy is "the only Hypothesis that Can explicate the *Phenomena,* or at lest that does explicate them so well." A number of science scholars have argued that Boyle's scientific method insured that his hypotheses were proven in this strong sense. To make just such a point, Rose-Mary Sargent maintains that Boyle adopted a method that she dubs "the concurrence of probabilities," borrowed in part from legal practice; according to this method, a conclusion was held to be rationally justified "only when all of the relevant and procurable evidence produced a 'concurrence of probabilities' in its favor and there was no evidence to the contrary." Here the term "probable" is taken to mean "worthy of approbation" and does not have a quantitative meaning expressed by degrees of likelihood. The method is meant to provide a moral demonstration that is compelling. Therefore, Sargent says, "If there is a concurrence of probabilities where all of the evidence favors a conclusion, then in order to be rational, one must assent to its truth." To show that all of the evidence favors a conclusion, one must perform tests to ensure that no phenomena are inconsistent with the hypothesis and all other hypotheses are eliminated.[2] According to Sargent, Boyle had two methodological strategies for such experimental tests: repetition and variation. One experiment is not enough; experiments should be repeated and the circumstances of their production should be varied, to ensure that the results are reliable as evidence for hypotheses. And finally, Boyle's method includes the demand for a mechanical causal explanation; to explain a property or quality of a body, one must describe the arrangement of its constituent corpuscles. Here "method" refers primarily to the type of causal explanation acceptable—a mechanical explanation in terms of the arrangement of corpuscles—and not to rules for the relations among evidence and hypotheses (Sargent, *Diffident Naturalist,* 83 and 176; cf. chapter 4, esp. 97–98).

The claim that air has spring and weight meets the demands of Boyle's method for determining matters of fact, according to Sargent, for it is a mechanical hypothesis with enough evidence to command rational assent, i.e., it is proven by a concurrence of probabilities. She says,

The hundreds of experiments reported in the first edition [of *New Experiments*], together with the additional experimental reports compiled in the second edition and two subsequent "continuations," all of which Boyle maintained, rightly or wrongly, could be explicated by the causal notion of a spring, allowed him to conclude that the air's spring could be accepted as a factual claim. . . . There were problems with all of them, as he freely acknowledged in his conclusion. But taken together, his experiments provided a set of collateral arguments and thus produced a moral demonstration.

Collateral arguments are none of them alone sufficient to prove a conclusion, but taken together do "directly prove" it (*Diffident Naturalist*, 134 and 134n).

The method of concurrence of probabilities that Sargent attributes to Boyle functions much like an early logic of science: the scientist must test and eliminate alternative hypotheses until one remains that agrees with all the phenomena of nature, not just the data it is framed to explicate. Since it is not possible to interrogate all the phenomena of nature, he must procure relevant evidence for the hypothesis by repeated and varied experimental trials. To be rational, he must assent to the hypothesis favored by the combined evidence from these trials, even if each separate trial is a bit problematic.

Sargent is among many Boyle scholars who have attributed a method to him. As we saw in chapter 2, Conant understands him to follow a version of the hypothetico-deductive method. Others have made him a Cartesian rationalist or a Baconian empiricist, and still others find him somewhere in between. For example, Laudan argues that Boyle had "a coherent and sophisticated view of scientific method" combining Bacon's experimentalism with Descartes's view in the *Principles* that scientists can make hypotheses about atoms, but cannot know anything about them with certainty.[3]

Laudan explicates Boyle's method as follows:

The scientist conducts wide-scale experimentation to determine the "divers effects of nature" [i.e., matters of fact]. He next suggests a hypothesis to explain what has been observed. The first hypotheses should be fairly low-level generalizations about the "immediate causes of the phenomena." Then, "ascending in the scale of causes," he arrives ultimately at the most general hypotheses, which concern the "more catholick and primary causes of things." At each level, the scientist checks to see if the hypothesis conforms to the corpuscular doctrine. If so, he tests the hypothesis against the entries in all his tables and against the other known laws of nature. If it is falsified he rejects it, if not, he continues to maintain it. (Laudan, "Metaphor," 90–91, quoting *W*, II, 37)

Did Boyle have "a method"? Reading across the many reports he gives of the observations and experiments upon which he based various conclusions, we find work reminiscent of the hypothetico-deductive method, of simple induction, and of a pragmatic appeal to what works. We also find, over and over again, the claim that a particular hypothesis is compatible with, "agrees rarely-well with," the data. Thus, it appears that Boyle has many methods. Sargent is forced to admit as much when she recognizes that Boyle appeals to usefulness and that the hypothetico-deductive account of his method is partially accurate. She remarks,

No one warrant is universally applicable. At different stages of inquiry, different warrants will be appropriate. . . . In some contexts rational inference and the testing of theoretical consequences may be required. At other times, it may be necessary to devote more effort to the collection of observations or to the performance of exploratory experiments. It is not possible to decide which methodological strategy will work best without taking account of the theoretical and technological context surrounding a particular area of study. (Sargent, *Diffident Naturalist*, 210, 211)

To cover all the different methods Boyle used, the concurrence of probabilities has become very broad, indeed.

Nevertheless, Sargent, Conant, Laudan, and others have each attributed to Boyle a method of proof according to which hypotheses are justified by their being mechanistic, by a particular relationship to experimental and observational evidence (each scholar specifies a different relationship), and by their coherence with other hypotheses, laws, and theories also justified on the basis of the evidence. Aside from problems with the demand for mechanistic accounts, none of the methods attributed to Boyle works in the way these scholars claim, for exactly when and how an experiment provides evidence for a hypothesis is not obvious. The history of twentieth-century philosophy of science shows how difficult it is to find a deductive relationship between hypotheses and evidence, or even a theory for assigning statistical probability to hypotheses in light of the relevant evidence. And as Richard Rudner and Carl Hempel argued many years ago, even if the scientist could assign probability in this way, something more is needed before he can rationally adopt a hypothesis. He must still have "adequate criteria for the rational acceptability of a hypothesis," taking account not only of its probability, in light of the relevant evidence, "but also of the values attached to avoiding the mistakes of accepting the hypothesis when it is, in fact, false; or of rejecting it when it is true."[4] The latter represents Rudner's insight, briefly, that once he has determined the probability of a hypothesis, the scientist must still decide to accept it, and his decision will take account of how important it is not to make a mistake, e.g., when scientists at a pharmaceutical firm decide that the probability of a drug's being both safe and effective is high enough to accept the hypothesis that "the drug is safe and effective."[5] (Hempel's recognition of this point led him to make the pragmatic turn and adopt epistemic values including the hypothesis's being probably true, rich in informational content, closely fitting experimental data, and having predictive power, simplicity, and compatibility with established theories in related fields [Hempel, "Induction"]. We will return to the role of values in scientific decisions below.)

The "concurrence of probabilities" Sargent attributes to Boyle has, of course, nowhere near the precision that methods using statistical probability would. Nothing in the method as Sargent describes it (or in Boyle's texts) tells Boyle when the evidence is sufficient to justify a hypothesis: for example, when the experiments he presented as compatible with the air's spring proved that hypothesis. They were certainly numerous; whether they were sufficiently numerous and sufficiently varied depends, of course, upon one's judgment, since the method is also silent on this point. Though he urged his readers to do so, Boyle does not report that he or anyone else repeated very many of his experiments, and he remarks in the conclusion to *New Experi-*

ments that he had not the leisure to repeat them, though he clearly repeated some. But in any case, his failure to repeat them certainly never prevented his drawing conclusions from them (*W*, I, 116, but see I, 90). And whether numerous experiments performed with the air-pump provided enough variation, i.e., whether more should have been performed using other instruments and whether the variations among experiments performed with the pump were too slight, is a matter of good technical judgment. But what counted as good technical judgment was precisely at issue between Boyle and those with alternative hypotheses. To address this problem, Sargent has recourse to the notion of expertise. An expert, "one who had developed his reason by experience in a specific area and was thus the most qualified to judge in that area," knew when the evidence is sufficient (Sargent, *Diffident Naturalist,* 50). Boyle was an expert; therefore, he knew when the evidence proved the hypothesis that the air has spring.

Of course, Boyle was an expert; but what this meant in practice we must learn by studying the many different ways in which he decided when the evidence favored a hypothesis. The concurrence of probabilities can be made to cover all these cases only at the cost of being too vague to be useful in determining, for example, when he had performed a sufficient number of sufficiently varied and repeated experiments.[6] We might note that Linus was also an expert in a different research program.

On the other hand, Shapin and Schaffer argue that Boyle had no detailed method for proving his hypotheses: "Boyle did not detail the steps by which he moved from matters of fact to their explanation. He did not, for example, say in what ways the air's 'elastical power' had been 'evinced' and established; he merely announced that this had been accomplished."[7] They urge us to attend to Boyle's actual practice, what Boyle himself does, rather than to what he says to do, to find out how he moves from matters of fact to hypotheses. Shapin and Schaffer argue that, although we believe Boyle was right about the best way to produce knowledge of the natural world, right to believe that matter is composed of inert atoms, and right again to believe that his vacuum pump removed nearly all the air from a glass globe, we should not analyze the dispute between Boyle and Hobbes (or, we might add, between Boyle and Linus) in our current, epistemic terms. Moreover, they refuse to argue that Boyle's program was historically successful because it was and is "true," "more rational," "more adequate," or "more objective," while Hobbes's (or Linus's) program failed because it was and is "false," "less rational," etc. Instead, they use the same types of cause to explain both true and false beliefs; thus, they argue that both Hobbes's and Boyle's programs were credible and they give the same sort of naturalistic explanation of each, that is, of Boyle's success and Hobbes's failure. We should not expect to find "a method" that was the source of Boyle's success; instead, historical investigation reveals how important religious and political concerns were to Boyle's success:

> We now consider the issues that bore on the way Hobbes's and Boyle's schemes were assessed in the 1660s. This demands an outline of the political and ecclesiastical context of the Restoration. The crisis of the Restoration settlement made proposals for a means of guaranteeing assent extremely urgent. We explore the importance of conscience and belief in the intellectual politics of the 1660s. The experience of the War and the Republic showed that disputed knowledge produced civil strife. It did not seem at all clear that *any* form of knowledge could produce social harmony. Yet this was just what the experimenters and their propagandists did claim. Furthermore, the restored régime concentrated upon means of preventing a relapse into anarchy through the discipline it attempted to exercise over the production and dissemination of knowledge. *These political considerations were constituents of the evaluation of rival natural philosophical programmes.* (Shapin and Schaffer, *Leviathan*, 283; final emphasis mine)

According to Shapin and Schaffer, the natural philosophies of Boyle and Hobbes were, in part, solutions to the political unrest that brought about the English Civil War and that later threatened the regime of Charles II. The success of Boyle's science depended upon its political success.[8] Nor were these political considerations unusual for scientists of the seventeenth century; instead, as Shapin notes,

> Seventeenth and eighteenth century moral uses of nature were not the "scientistic" extrapolation of esoteric natural scientific findings onto social problems; the moral and social uses of na-

ture were essential considerations in the evaluations historical
actors made of various theories, models, metaphysics and statements of fact.[9]

Boyle distinguished the criteria both for good and for excellent hypotheses. But there are problems when we treat these criteria as if
they constitute "a method" and try to use them to decide between
the hypothesis of the air's spring and Linus's funicular hypothesis.
Many philosophers have pointed out that criteria in such lists have
no precise definitions.[10] For example, simplicity has proven notoriously difficult to define; depending on the case, the simplest hypothesis may, *inter alia,* propose the fewest ontological entities or be mathematically elegant (a concept itself in need of definition). And whether
a hypothesis is predictive depends upon factors such as the margin of
error. This does not mean the criteria are invalid, but it does mean
that we need to examine each appeal to such criteria in order to understand their use.

Finally, it could be supposed that, while certain assumptions encouraged Boyle to pursue the mechanical philosophy, these assumptions held little attraction for his peers, who were swayed only by empirical considerations. Peer review, after all, ensures that the biases of
individual scientists are discovered and neutralized. We find, however, that assumptions about the support for radical religion and politics provided by animistic natural philosophies were widely shared
among Boyle's contemporaries.[11]

Feminist scholars of science have argued that while peer review can
be helpful in detecting individual bias, it is not as successful in detecting assumptions shared by a homogeneous scientific community.
Lynn Nelson shows us the connections between the gender politics of
scientists and the accounts of nature that science communities develop by turning to Quine's understanding of science as "a bridge
of our own making," yet a bridge which is constrained by experience, and to his view that common-sense, scientific, and philosophical theories are fundamentally connected and interdependent.[12]

Feminist science criticism, Nelson points out, indicates that directions of research and the content of theories often reflect such factors as scientists' experiences of sex/gender and politics, "not just in
cases where the charge of 'bad science' is appropriate . . . but in cases

where the charge is not appropriate." She uses the term "androcentric" (male-centered) to refer to the fact that in many scientific views, the language describing phenomena, the questions pursued, the models adopted, the interpretations made of data and observations, and the theoretical frameworks developed reflect the fact that, typically, results based on men's experiences, behaviors, and activities are generalized to women, or attempts are made to assimilate women into the androcentric framework. Nelson does not use the term "bias," since that term implies conscious manipulation of data and suggests that scientists could and should be disengaged observers of the world (*Who Knows*, 189ff.). It is implausible, she says, that androcentric research is underwritten by a conscious desire to justify men's domination of science or to reaffirm their own sexist views. Instead, the connection and interdependence of our theories can lead to androcentric work. Although some androcentric work appears unwarranted in isolation—studies, for example, that attempt to connect male and female hormones with hemispheric lateralization in the absence of evidence that there are sex differences in lateralization—such work may be strongly connected to other working hypotheses, studies, and research programs. Thus, although some feminists have criticized the pervasive and conscious borrowing of hypotheses and results among research programs and sciences investigating sex differences, Nelson argues that this borrowing is not a mark of flawed scientific procedure. Instead, scientists depend upon one another's work. "On the other hand," Nelson points out, "a *hundred* research projects attempting to establish or explain sex differences . . . will not add up to *one larger good* research project if each incorporates a wrongheaded assumption." And feminist criticism of science indicates that various wrongheaded assumptions about sex differences, divisions of labor by sex, and so on, are shared by many sciences and research programs (*Who Knows*, 200–202).

Nelson argues for the view that communities are the primary epistemic agents, constructing and acquiring knowledge, and she notes that the impetus for the traditional empiricist commitment to individuals as the primary epistemic agents came from the empiricist demand that our theories be tied to the world. But the demise of foundationalism (the view that untheorized observations or experiences tie our knowledge directly to the world) requires a more complicated epistemology. Instead of a direct connection between theories and the world, we find a connection between a system of going theories and our experiences of the world. On Nelson's account, we experience the world "through the conceptual scheme, the theory of nature, we begin learning as we learn language." Our experiences are possible not only because we have sensory receptors, but also because we are able to learn public theory. These always already theorized experiences, together with our going theories, constitute the constraints on what it is reasonable to believe. We see that public theories are necessary, not only in order to have experiences, but also to have knowledge, because individuals do not determine what constitutes evidence for a claim. Instead, what constitutes evidence for a claim is determined by the standards a community accepts "concomitantly with constructing, adopting, and refining theories"—scientific theories as well as common-sense theories (*Who Knows*, 276–277).

Thus, on Nelson's account, gender politics sometimes enter scientific theorizing because sex/gender assumptions are included in our common-sense beliefs, and our common-sense theories and scientific theories are interdependent. And given that our scientific theories and research programs depend upon one another, borrowing hypotheses and results from one another, we should not be surprised to find sex/gender assumptions in many of them. Thus Nelson's social empiricism shows that androcentric assumptions are sometimes found in scientific knowledge, not because of any individual scientist's bias, but because many scientists collectively share both common-sense and scientific androcentric assumptions. To help solve this problem, Nelson suggests a sophisticated type of self-reflection among scientists; other feminist scholars of science also suggest that self-reflexivity and diversity among scientists would help and should therefore be built into scientific method.[13]

19

Good Science

For many people, there is something deeply wrong with a book that says science is not always value-free. "Everyone knows" that good science is value-free and that if values influence scientific work, the result is bad science. So this book must be wrong about Boyle. Since his work was good, he cannot have let his religious and political views influence his choice of theories. Or else the book is wrong because it implies without saying so that, since Boyle's work leading to the Gas Law was influenced by his religious and political views, it was bad science after all.

How could Boyle's ideas about proper religious, social, and political arrangements, including gender arrangements, influence his choice of mechanism and so his laboratory work, his collection and interpretation of data, and still allow him to produce good scientific work? In this chapter, we will see how good science can be influenced by gender and class politics and still be good science. But first we need to distinguish two kinds of values.

Philosophers distinguish many kinds of values in connection with science, but two are relevant to our argument. One kind is said to be acceptable for science; these are called constitutive values (or epistemic values or cognitive virtues). Boyle referred to them as "Requisites for a Good [or "Excellent"] Hypothesis." The list varies among philosophers but usually includes the theory's being probably true, closely fitting experimental data, and having informational content, predictive power, simplicity, compatibility with established theories in related fields, refutability, internal consistency, etc. Scientists use such criteria in evaluating their hypotheses and theories. The other kind of values is called contextual (or non-epistemic or non-cognitive), and it includes political, religious, economic, and any other values not considered constitutive.[1] Although we begin this discussion by making this distinction between two kinds of values, we will end by seeing that in the actual practice of science, there is no clear distinction.

One of the problems in dealing with the connection between science and contextual values is stating precisely what the nature of the connection is alleged to be. Longino distinguishes five ways in which contextual values can affect good scientific work; they can

1. affect practices that bear on the integrity of science, as when the desire for profit leads a scientist-entrepreneur to present his results first at a press conference rather than in a professional journal or at a professional conference;
2. determine which questions are asked and which ignored;
3. affect the description of data, that is, value-laden terms may be employed in the description of experimental or observational data and may influence the selection of data or phenomena to be investigated;
4. be expressed in or motivate the background assumptions facilitating inferences in specific areas of inquiry; or
5. be expressed in or motivate acceptance of global, framework-like assumptions that determine the character of research in an entire field. (Longino, *Science*, 86)

Our study of Boyle's work shows that contextual values influenced his choice to pursue a mechanistic research program (way 5) and so influenced his interpretation of data, e.g., the data confirming Boyle's Law (way 4).

We need a model of scientific thinking that shows us how it is possible for scientific work to be influenced by contextual values and yet still be good science. Such a model allows us to see that background assumptions about proper religious, class, and gender arrangements, along with assumptions about the natural world, constrained the choices Boyle made about how to pursue his experimental work. And, most importantly, such a model allows us to see that Boyle's work was, nevertheless, excellent: predictive, fruitful, and with good explanatory power. There are several models or philosophies of science available that will serve these purposes. I have found Mary Hesse's network model of scientific theories useful for pursuing the gender politics in seventeenth-century science.[2] Hesse understands a scientific theory as a system of laws which has a very complex relation to nature.[3] When a scientist establishes a law, Hesse argues, he classifies phenomena on the basis of resemblances among them. Any scientist is, then, constantly faced with decisions as to whether two things are similar enough to be classed together. But since phenomena are similar in some respects and different in others, the question becomes "Which respects are more important, the similar ones or the dissimilar ones?" When the data—here, observations of the respects in which phenomena do and do not resemble one another—are all in, decisions must be made about which are significant. This is a fundamental case of "interpreting the data." Data alone, observations alone, do not determine a law or generalization; for example, we observe that whales swim in the water and so are like fish; but we also observe that they are live-bearing like mammals. Are they fish or mammals? Because similarity is not transitive, a decision must be made on grounds other than observed similarity. That is, *b* may resemble both *a* and *c,* but *a* and *c* do not therefore resemble one another; how, then, should we classify *b*? As an *a* or as a *c*? Since any decision here is underdetermined by the data, it has to be determined on other grounds.

One criterion at work in such a case is logical coherence throughout the system; i.e., laws must not contradict one another and neither should observations. However, we cannot claim that coherence alone is sufficient to account for theory production. Scientists do not always decide between conflicting observations on the grounds that one generalization provides coherence with the greatest number of other generalizations. The problematic generalization may instead trigger the decision that most of the generalizations in the theory are wrong.

At this point, the mainstream philosophers who adopt a network model argue that scientists either do or should have recourse to constitutive values. Scientists hold or should hold certain assumptions about what constitutes good systems of laws or "good theories." Just so, philosophers have argued, the assumption that good theories are "conservative" or "simple" guides the scientist to make the decision that conserves most of what has been held true in the past, or the one that makes the system simpler. We have seen a list of constitutive values above. Hesse refers to these values as "coherence conditions" and argues that they include assumptions such as the goodness of symmetry and of certain analogies, models, and so on (Hesse, *Inference,* 52). However, feminists, as critical scholars of science, want to know, not what mainstream philosophers of science think scientists should do, but what scientists actually do. That is, we need to see what assumptions scientists actually hold to when they decide between conflicting generalizations. The feminist working hypothesis is that the assumptions guiding classificatory decisions may include assumptions about proper gender and class arrangements.[4]

Thus, feminist concerns lead to an extension of the network model by recognizing gender (as well as class, race, and other) assumptions as "coherence conditions." Symmetry, favored analogies, and models, like the traditional constitutive values, are still "technical" considerations, suitable for an internalist account of scientific theory production. But an extended model shows us how gender and class assumptions can influence the construction of scientific theories. At least for those cases in which a particular generalization remains underdetermined by the data, the decision as to which generalization to adopt must be based on other grounds than simple observation. Because all the generalizations in a system are logically interrelated, the adoption of one of a pair of conflicting generalizations will have repercussions throughout the system and throughout related systems. The assumption of technical coherence conditions (constitutive values) can determine which repercussions are desirable, but so can the assumption of some other principle, for instance that male behavior is the norm, that male behavior is crucial to evolution, that hierarchies are functional, that hierarchical models are better than non-hierarchical ones, or that women should remain in a secondary social position. The suggestion here is that studies of scientific work should look carefully at the constraints affecting the choices scientists make between conflicting generalizations. We sometimes find that in actual scientific work, both constitutive and contextual values function as coherence conditions. On a network model each generalization in the system is—at any given time, though not at all times—corrigible, i.e., any generalization can be modified, so there is theoretically nothing to prevent us from discovering that even the most innocent-seeming choice is constrained ultimately by a gender or class assumption.

The model allows us to see that, theoretically, each decision is underdetermined by the data; in theory, i.e., according to this model, it is always possible to hold to a single decision and redistribute truth-values throughout the system as necessary to make all the generalizations consistent. But, historically, this rarely happens. Once scientists have made a number of decisions about how to interpret data, generalizations are held true and become statements of fact. And sometimes scientists hold on to low-level facts while high-level principles and facts are in doubt. Boyle was working at a time when facts about suction pumps, the behavior of liquids sucked into reeds, etc. were held firm while principles of explanation and even low-level generalizations were up for grabs. Thus, Boyle and Linus agreed about many of the data but disagreed over fundamental principles and explanatory hypotheses such as the weight and spring of the air.

But once some principles are in place and some decisions made about data, it is hard to remember that, theoretically, one could give up even "obvious" facts and fundamental principles and redistribute truth values throughout one's system; one could hold new principles and facts to be true and old ones (inconsistent with the new) false. When a number of such decisions have been made, conservatism comes into play and it appears foolish even to think of giving up the whole system. This was probably how Linus felt. Nevertheless, occasionally one does, and Boyle did. But, sure enough, once Boyle committed himself to a set of principles and facts, and found that he could interpret yet more data in ways consistent with them, infelicities such as the cohering marbles and the problematic nature of gravity did not sway his commitment.

The flexibility of any system of generalizations means that choices among generalizations can be made that allow at once a reasonable degree of empirical adequacy, coherence, and fruitfulness, as well as simplicity, faithfulness to preferred analogies or models, and the maintenance of religious, class, or gender assumptions. Thus, the model makes it clear that even good scientific theories, good by all the traditional criteria, can be influenced by value assumptions constraining the distribution of truth values throughout the system. The flexibility of theoretical systems also allows the possibility of new and different theory constructions. Theories could be constructed that base classificatory decisions on assumptions of social equality instead of inequality. The maintenance of an assumption, such as the assumption that women and men should be allowed equal social positions, would have repercussions throughout the classificatory system and would still allow, just as inegalitarian assumptions have, some degree of empirical adequacy, coherence, and fruitfulness, along with simplicity, faithfulness to preferred analogies or models, and so on.[5]

Finally, we should note that while the network model shows that some coherence condition(s) must be used by scientists, none, including contextual values such as those mentioned above, operates necessarily or always or even most of the time. Science scholars must look at each case to determine which coherence conditions scientists employed.

Hesse's network model is not the only one allowing us to see whether and how contextual values influence scientific decision making, including good scientific decision making.[6] Longino argues that scientific inquiry is group work in which the acceptance of models and theories depends on an interplay of data and background assumptions, and she shows that contextual values can enter the confirmation of models and theories through background assumptions. Countering the arguments of positivists and others, Longino holds that there is no unique or intrinsic evidential relation; rather, the connections or regularities we appeal to in assessing evidential relations are connections or regularities from some point of view and are always subject to change. That is, people take e as evidence for h when they believe e belongs to a class of things which are related (causally or through class membership) to the class of things h belongs to. The objects, events, and states of affairs providing evidence for hypotheses do not carry labels showing what they are evidence for. Instead, how one determines evidential relevance depends on one's background beliefs or assumptions, and these background assumptions can include contextual values.

But because what counts as evidence for a hypothesis and how strongly it supports the hypothesis are relative to the background beliefs one holds, the objectivity assumed to characterize scientific inquiry appears to be in jeopardy. To solve this problem, Longino argues that the production of scientific knowledge involves such social activities as repeating the experiments of others and subjecting their work to peer review. And until the results are absorbed into the ongoing work of the scientific community—as evidenced by the citation, use, and modification of the work—the work cannot be said to constitute "scientific knowledge." Objectivity is, thus, a result of the social nature of scientific work (*Science*, 68, 71ff). Social activities such as peer review and repeating experiments can weed out contextual values serving as background assumptions, but if the contextual values are widely shared among scientists and others, they can remain unnoticed.

At this point it is useful to distinguish two senses of "good" as applied to scientific work: instrumental and moral. Instrumentally good science, as we have seen, has empirical adequacy and satisfies some list of constitutive values; but because there are no precise criteria for the application of any of these concepts, the scientist or the relevant scientific community must decide when a theory has enough empirical adequacy, is general enough, predictive enough, etc. to be accepted. It can then be deemed a good theory. With our distinction in mind, we want to say that it is deemed a good instrumental theory, though it is not clear that this is what scientists "really mean" when they dub something "good work." As we noted above, theories are flexible, allowing many choices among the generalizations that compose them, so that the system as a whole can maintain benign or malignant contextual values and a reasonable degree of empirical adequacy, as well as reasonable degrees of other constitutive values. If it maintains a benign contextual value, we might deem it a morally good theory or decide that it is morally unobjectionable. In fact, if it maintains a common value, one that is part of the moral status quo, it is likely to pass as morally neutral. And this is the second and, as I think, proper sense of the phrase "good science," viz., morally neutral science. The value-neutrality thesis is embedded deeply enough in scientific culture that calling a theory "good" because it maintains a widely accepted moral value is an improper sense of the phrase "good work." Thus, in common scientific parlance, "good science" means instrumentally good science that is morally value-neutral.

The assumption is that all good science is contextually value-neutral, from which it would follow that science influenced by any contextual values, good or bad, is instrumentally bad science (and dubbed "biased"). But only empirical investigation which does not beg the question by assuming at the outset that good science is contextually value-neutral can determine whether cases of good science are or are not neutral. Studies that are methodologically agnostic about the contextual value-neutrality of good science have produced plausible case histories revealing the influence of moral, religious, and other contextual values upon good science.[7] Thus, the inference that any science influenced by contextual values is instrumentally bad science is either empirically premature or question-begging. Moreover, Robert Boyle's work leading to the Gas Law presents us with a case of good science which was not free of contextual values.

Conclusion

We have found that Robert Boyle was interested in gender issues in the mid- to late 1640s, the time of his early reading and thinking about the many physical theories competing for attention. As women took up the pen to petition and took to the streets to protest, Boyle was writing essays, letters, and a romance describing an ideal woman, one who breastfed her babies, wore no make-up, was modest in dress and demeanor, and was chaste and pious. This woman, we found, made a good foil for Boyle's ideal experimental scientist, a professional man who remained chaste and modest in his own way, devoting his life to the service of God through his work revealing the wonders of God's universe. Though this man could be tempted by immodest women, he refused temptation and turned his energies to science.

When we tell the story of Boyle's work leading to the Gas Law in a traditional way, we find no connection between his discovery and his thinking about gender. To find the connection, we need to look at Boyle's work in its social and political context. Boyle came to England in 1644, in the midst of the Civil War. It was a chaotic time during which the king's power waned, and with it the power of the Anglican Church to enforce censorship and conformity. Thus, nonconformist sectaries preached and wrote of enthusiastic religious ideas, ideas that reinforced politically radical notions that included greater class and gender equality. Radical enthusiasm found expression in many bizarre political goals, but also in many goals that we take for granted: radicals wanted, among many other things, religious freedom, an end to tithes, manhood suffrage, the right to petition for redress of grievances, and the consent of Parliament to war and peace. Hoped-for economic reforms included lifting the excise from commodities necessary to the poor, the abolition of monopolies, the simplification and cheapening of legal procedure, and a change in the practice of imprisonment for debt. Many also argued for the equal distribution of wealth.

Women petitioned for these and other things, even though they had no legal right to do so. They called for better economic conditions, complaining of the decay in trade, the high price of food, lack of work, high taxes, and tithes. They argued for the establishment of a state religion and for an end to the civil wars, and women who supported the Levellers' cause petitioned for the release of Leveller leaders and against Parliament's arbitrary exercise of power. In 1649, some ten thousand women petitioned, claiming political equality with men.

The ideas expressed by these men and women would, if carried out, have "turned the world upside down." And when the king was beheaded in 1649, many thought that time had come. But as hopes for a stable republic dimmed and a new king returned from abroad and re-established order, "enthusiasm" made a convenient scapegoat for the turbulent interregnum.

We found that many radicals based their political aspirations upon a hylozoic, animistic metaphysics particularly associated with Hermes Trismegistus and having its roots in the Neo-Platonic natural-magic tradition. This tradition ran through the Middle Ages into the modern period; it was taken up and worked out by thinkers from Ficino through Paracelsus and Campanella and found expression in Helmont's extraordinary blend of animism with atomism and quantitative experimentalism. Since the mid-thirteenth century, animism and the natural-magic tradition had been associated with political radicalism, and they were associated with radical enthusiasm in early modern England. But because the thinkers who took up the animistic natural-magic tradition used it to provide alternatives to Aristotelian physical and biological theories, the tradition was also associated with new possibilities in medicine, biology, chemistry, and physics.

Boyle himself favored religious freedom within limits, but he viewed sectaries as "upstarts" and looked forward to their destruction. He certainly did not favor torturing men for their religious opinions, but he also associated radical enthusiasm with animism, which he argued strongly against. Although he gave many purely theological objections to animism, many of his arguments blended theological arguments with objections to physical and biological theories and hypotheses that attributed consciousness and thought to natural things like water or to Nature "herself."

Scholars have argued that Boyle rejected active principles in nature and took up mechanism in the mid-1650s because the belief that natural objects are self-moving was associated with radical enthusiasm. This argument has been bolstered by indications that Boyle was himself interested in the natural-magic tradition and that until the mid-1650s he held some version of hylozoism, particularly as evinced in Helmont's physical and biological theories. Recent scholarship has confirmed that Boyle was very interested in magic, and scholars have argued that he retained some belief in active principles throughout his life. If true, this argument helps us see that Boyle did not "convert" from hylozoism to mechanism in the mid-1650s.

We agree that Boyle made no such "conversion" and, for the sake of argument, we have granted that Boyle used active principles to explain some phenomena long after the mid-seventeenth century. But our argument does not hang on whether Boyle was ever a complete mechanist. His own words make clear that he opposed animism. He objects, he says, to the commonly received notion of Nature as having intelligence, wisdom, and providence and points out that "excessive veneration" for her can lead men into "extravagant and sacrilegious errors," including pantheism, in which God is "so united to matter, as to animate it like the heathen's mundane soul," and belief in a soul of the world. Indeed, "some of our late infidels pretend to be great discoverers of new light in this affair." On the other hand, he is happy to point out, mechanism "may keep many, that were wont, or are inclined, to have an excessive veneration for what they call nature, from running, or being seduced, into those extravagant and sacrilegious errors."

Boyle argues that this pernicious belief in an animated nature lies behind Aristotelian and other hylozoic physics. For example, he objects to the hylozoic interpretation of phenomena such as suction of water through a reed: if an Aristotelian sucks liquid into his mouth through a reed, "he will readily tell you, that the suction drawing the air out of the cavity of the reed, the water must necessarily succeed in the place deserted by the air, to prevent a vacuity abhorred by nature" (*W*, II, 37). The problem with this account, Boyle says, is its assumption of a World Soul: "to say, that the ascent of the water . . . proceeds from nature's detestation of a vacuity, supposes that there is a kind of *anima mundi,* furnished with various passions" (*W,* II, 37–38). This account also assumes that water is conscious.

Against such an account, Boyle took up the view that the earth is surrounded by a sea of air which has weight and spring. This weight and spring account for why, when the Aristotelian sucks on the reed, water is drawn up into it, for when air is sucked from the reed, water is forced up into it by the weight and spring of the air leaning upon the surface of the water. Similar things happen in many experiments with the pneumatical engine, e.g., when a Torricelli apparatus is placed within the receiver and air is pumped out, the mercury in the tube falls because the weight and spring of the surrounding air are reduced.

Franciscus Linus's alternative explanation was not simply that the mercury "follows" the air as it is exhausted from the receiver in order to prevent a vacuum; he argued that as the air is exhausted, it thins into a membrane adhering to the surface of the mercury in the basin and draws it up, thereby drawing the mercury down in the tube. In the same way, Linus argued that a membrane of mercury forms during the original Torricelli experiment and holds the mercury up to a height of 29½ inches. Linus, of course, was not a radical, and it is very unlikely that he thought his Aristotelian physical principles supported radical Protestant sectaries. (He was a Jesuit, and as such represented a different threat to Anglican stability in England.) But even though he was not what we have called a "lively animist," nevertheless, when Boyle attacked Linus's views, he treated Linus as defending the animism to which he had such profound objections. And Boyle tells us that while he was working to defeat Linus's alternative hypothesis, he discovered the inverse relationship between the volume of air and its weight and spring ($k = pv$ or $p_1v_1 = p_2v_2$).

Through our analysis of Boyle's dispute with Linus, we found that both theories could account reasonably well for the experimental data available at that time. (Linus's funicular hypothesis could even have yielded Linus's Law of Funiculi, $k = hw$ or $h_1w_1 = h_2w_2$.) But Boyle was defending his mechanistic accounts against Linus, for he was already committed to a mechanistic research program, the mechanical philosophy. Given the empirical promise of mechanism and his strong objections to animism, associated as it was with radical class and gender politics, Boyle chose to pursue a mechanistic research program, and found that the hypothesis that the air has weight and spring accounted well, though certainly not in all cases, for his experimental data. But we also found that, given his commitment to (a version of) animism and the empirical successes of scholastic science (as well as its strong coherence with other established theories in physics, biology, and medicine as well as in psychology, theology, and philosophy), and given that his funicular hypothesis accounted reasonably well, though certainly not in all cases, for Boyle's data, it was reasonable for Linus to stick to his guns.

Boyle, we found, was aware that hypotheses are underdetermined by data, and there are indications that he was aware that Linus's hypothesis accounted reasonably well for his data. But he felt that Linus had given him no reason to abandon mechanism; thus, he charged that Linus's hylozoic principles in general and his funicular hypothesis in particular were "unintelligible," i.e., non-mechanical, as well as precarious and insufficient to account for the data. He remained convinced that his experiments confirmed his hypothesis—and he was reasonable in doing so. But his commitment was based not only on the empirical adequacy of his hypothesis, but also on its religious and political meaning, including its implications for gender politics.

If we pay attention only to what Boyle and others wrote about his experimental work, we get only half the story behind his pursuit of a mechanistic research program. We have found that he decided early on—probably in the mid-1650s—to pursue the mechanical philosophy, persuaded in this choice not by the mysterious insight of genius or by foreknowledge of its truth, but by both his objections to animism, with its train of ideological associations, and the empirical promise and successes of mechanism (and despite the embarrassment of failed experiments, such as the cohering marbles).

In response to the strongly felt objection that because Boyle's work was and is a paradigm of good science it cannot have been influenced by religious and political values, we set out Hesse's network model of scientific theories. This model is one of many that allows us to see how both constitutive and contextual values work as coherence conditions, serving on occasion along with empirical adequacy to direct the distribution of truth values throughout a system of hypotheses. Thus, a system of hypotheses (or a theory, or a set of theories) can exemplify good science—it can be at once empirically adequate, generalizable, and of reasonable scope, it can fit a favored model, etc.—and still reflect contextual values such as those we have examined in this study of Boyle's work. In this way, both empirical and ideological considerations convinced Boyle to accept mechanistic principles; and once adopted, these served as guides to the kind of hypotheses to put forward, the experiments to perform, and the interpretation of resulting data, including especially experiments leading to Boyle's Law of Gases.

Twentieth-century scientists and philosophers of science argue that good science is neutral among competing moral and political values. And of course, when scientific method includes values such as peer approval, the value is taken as a constitutive one. More controversially, when scientific work includes political neutrality or leaves current assumptions unchallenged, it looks neutral but can be seen as favoring the status quo. Boyle's case shows us that the assumption that contextual values cause bad science, or conversely, the assumption that good science is value-neutral, is false. These assumptions are very deep, so our study will be taken ipso facto to debunk Boyle, i.e., as trying to show that he was a bad, biased scientist. We know that he was a very good scientist and deserves to be remembered for his work on the Gas Law. But his work was not neutral among contextual values.

Notes

INTRODUCTION

1. See Alison Wylie et al., "Philosophical Feminism: A Bibliographic Guide to Critiques of Science," *Resources for Feminist Research* 19:2 (1990), 2–36; Londa Schiebinger, "The History and Philosophy of Women in Science: A Review Essay," *Signs* 12:2 (1987), 305–332, and, more recently, her *Has Feminism Changed Science* (Cambridge and London: Harvard University Press, 1999), esp. Part III. See also Hilary Rose, "Beyond Masculinist Realities: A Feminist Epistemology for the Sciences," in *Feminist Approaches to Science,* ed. Ruth Bleier (New York and Oxford: Pergamon, 1986), 54–76; Carolyn Merchant, "Isis' Consciousness Raised," *Isis* 73:268 (1982), 398–409.

1. NOW WE SEE IT

1. Preface to his story "The Martyrdom of Theodora and Didymus," in *The Works of the Honorable Robert Boyle,* ed. Thomas Birch, vol. V (London: J. & F. Rivington, 1772), 259–260. All future references to *The Works of the Honorable Robert Boyle* will be to this edition and will be cited in the text as *W.*

2. Letter from Katherine Ranelagh to Boyle, June 3 (no year given), *W,* VI, 522; letter from Boyle to Mrs. Hussey, June 6, 1648, *W,* VI, 236; letter from John Wallis to Boyle, July 17, 1669, *W,* VI, 458–460; and see *W,* I, cxxxviii.

3. Letter from Katherine Ranelagh to Boyle, October 12 (no year given), *W,* VI, 523.

4. Boyle Papers, vol. XXXVII, 159. I thank the president and board of the Royal Society of London, where Boyle's unpublished manuscripts are located, for permission to quote from this material. Future references will be cited in the text as BP and will include volume and page numbers where available.

5. James R. Jacob, *Robert Boyle and the English Revolution: A Study in Social and Intellectual Change* (New York: Burt Franklin, 1977), 81.

6. This argument also appears in Boyle's "The Martyrdom of Theodora and Didymus." When Theodora's friend Irene asks Theodora why she won't marry Didymus, Theodora, whom modern readers have taken to voice Boyle's sentiments, replies that if she were married, it would be her duty to worry about a near friend's danger; she would not be content to die because duty and inclination would, she worries, "fasten" her to life. Chastity allows one to serve God "more undistractedly and more entirely" and be "uninterruptedly employed in the direct contemplation and services" of a "sublime object." She doesn't see why anyone who doesn't need to "should enter into a relation that would make those distracting duties necessary" (*W, V,* 277).

er to a sister in Ireland, December 21, 1649, *W,* VI, 51; letter from e Ranelagh to Boyle, June 3 (no year given), *W,* I, 522 and *W,* I, 248. Evelyn to Boyle, September 29, 1659, *W,* VI, 401.

9. I am indebted to Shapin and Schaffer for pointing out the importance of modesty in the experimental production of facts. See Steven Shapin and Simon Schaffer, *Leviathan and the Air-Pump: Hobbes, Boyle, and the Experimental Life* (Princeton: Princeton University Press, 1985), chapter 2. Shapin and Schaffer discern three technologies used by Boyle to produce facts: material technology, including the spaces in which experiments were performed and the machines they were performed on; social technology, ways to determine who was eligible to produce knowledge and to bring about consensus and manage dissensus among those eligible; and literary technology, texts displaying the new subject of scientific activity and providing for, among other things, the mediate witnessing of his experiments. It is in the last, the literary technology, that we find Boyle's construction of the modest man of science.

Perhaps because they are not interested in the issue of gender, Shapin and Schaffer never discuss characteristics such as chastity or modesty as gender characteristics. They do not, therefore, mention the interpenetration of gender issues with those of style or, indeed, with any other technical matters constituting seventeenth-century experimental science.

10. Thomas Sprat, *History of the Royal Society,* ed. Jackson I. Cope and Harold Whitmore Jones (Saint Louis: Washington University Studies, 1958), 111.

11. T. J. Pinch, "Theoreticians and the Production of Experimental Anomaly: The Case of Solar Neutrinos," in *The Social Process of Scientific Investigation,* ed. Karin D. Knorr, Roger Krohn, and Richard Whitley (Dordrecht and Boston: D. Reidel Publishing Company, 1980), 77–106.

12. Pinch, "Theoreticians," 100, 93–94. Although the scientists whom Pinch interviewed cited Davis's modesty, others disagree and argue that the theoreticians were impressed not by Davis's modesty, but by his reliability, conscientiousness, and responsiveness to criticism (Jane Camerini, personal conversation, November 1997). It is, of course, not essential to our point that all scientists agree on this issue.

13. William Gouge, *Of Domesticall Duties* (Amsterdam and Norwood, N.J.: Walter J. Johnson, 1976; facsimile of the 1622 edition), 9.

14. Linda Woodbridge, *Women and the English Renaissance: Literature and the Nature of Womankind, 1540–1620* (Urbana and Chicago: University of Illinois Press, 1984), 159ff.

Boyle himself remarks that "Virtue is not ty'd to Fields or Breaches; tis the successful Combat with our Passions that is true Valor; that with our Enemys is so but as it is Symtome & the Effect of the other (in the minde)" (BP 195, quoted in Jacob, *Robert Boyle,* 62).

15. *Anatomy of Abuses,* Sig. [F5]-[F5]^v, and *A Sermon of Apparell,* 7, both quoted in Woodbridge, *Women,* 140, 142.

16. *Hic Mulier,* Sig. B2, emphasis mine, quoted in Woodbridge, *Women,* 146ff.

17. Sprat, *History,* 145. Margaret Cavendish was probably the first woman to attend a meeting. She apparently asked to come, was granted permission by the

president, but held out for an invitation from the Society. See Thomas Birch, *The History of the Royal Society of London,* vol. 2 (London, 1756), 175ff.

18. This is so despite his occasionally mentioning women as sources of information on topics of concern to him. Recently, Rose-Mary Sargent has argued against Shapin and Schaffer that Boyle's use of information given him by women means that women were considered credible witnesses of scientific work. See her *The Diffident Naturalist: Robert Boyle and the Philosophy of Experiment* (Chicago: University of Chicago Press, 1995), 152. I believe that there are two senses of the word "witness" operating here. Shapin and Schaffer took the word up from Boyle as a technical term meaning someone who provides authoritative corroboration that an experimental matter of fact is as claimed (paradigmatically, claimed by Boyle about what happened in the air-pump). See Shapin and Schaffer, *Leviathan,* esp. chapter 2. Sargent employs the term in a looser sense to mean, roughly, someone who provides testimony that such and such a thing happened. In none of the cases cited by Sargent can we say that the women are considered virtuosae in the way that, for example, Dr. Wallis is a virtuoso. Unfortunately, they are not the new women of science.

19. *W,* III, 265–266 and 205–206. Again, scholars have recently contested the claim that working-class folk were not witnesses to the production of facts. Sargent argues that Boyle was democratic in his production of facts: "each person was to be assessed for credibility based upon individual experience and competence. The reports of different classes were mingled together in Boyle's essays and histories" (*Diffident Naturalist,* 152). But if the workers who assisted Boyle were authoritative witnesses, one would expect them to be mentioned by name in the way that Dr. Wallis, Dr. Ward, and Mr. Wren are (e.g., *W,* I, 34). See Steven Shapin's *A Social History of Truth* (Chicago: University of Chicago Press, 1994), for extended treatment of this and related issues.

20. See *W,* I, 1 and 90; "This experiment was a few days after repeated, in the presence of those excellent and deservedly famous Mathematic professors, Dr. Wallis, Dr. Ward and Mr. Wren, who were pleased to honor it with their presence" (*W,* I, 34); "This relation of an experiment, which I afterwards showed to many virtuosi" (*W,* I, 410); "Of which some of the chiefest [of my experiments], and some of the most difficult, having been seen . . . by the Royal Society itself, or by inquisitive members of it, it will, I presume, be but a reasonable request, if the reader, that shall have the curiosity to try them over again, be desired not to be too hasty in distrusting the matters of fact, in case he should not be able at first to make everything succeed according to expectation" (*W,* II, 743). See also *W,* III, 184, 185, and 187.

2. NOW WE DON'T

1. Thomasio Campanella, *On the Sense and Feeling in All Things and on Magic,* in *Renaissance Philosophy,* vol. 1, ed. and trans. Arturo B. Fallico and Herman Shapiro (New York: The Modern Library, 1967), 364.

2. W. E. Knowles Middleton, "The Place of Torricelli in the History of the Barometer," *Isis* 54, pt. 1, no. 175 (1963), 12–13.

3. All quotations are from Aristotle's *Physics,* 213b–217b, trans. R. P. Hardie

and R. K. Gaye, in *The Basic Works of Aristotle*, ed. Richard McKeon (New York: Random House), 19.

4. "We see that numerous tracts are written in Poland on the vacuum in the glass tube, but nothing comes of it, and similarly in our experiments, of which even now we are multiplying new ones, but nevertheless we conclude that it is rarefied air, not vacuum" (letter to Hevelius, July 27, 1648, quoted in Charles Webster, "The Discovery of Boyle's Law, and the Concept of the Elasticity of Air in the Seventeenth Century," in *Archive for the History of the Exact Sciences* 2 [1965], 451).

5. Quoted in Webster, "Discovery," 458. I have changed spelling and punctuation for the ease of modern readers.

6. Shadwell quoted in Steven Shapin and Simon Schaffer, *Leviathan and the Air-Pump: Hobbes, Boyle, and the Experimental Life* (Princeton: Princeton University Press, 1985), 155. Samuel Butler, *The Elephant in the Moon*, quoted in Flora Masson, *Robert Boyle: A Biography* (London: Constable and Company, 1914), 251. Samuel Pepys, *The Diary of Samuel Pepys, M.A., F.R.S.*, ed. Henry B. Wheatley (London: George Bell and Sons, 1894), vol. 4, 27.

7. James Bryant Conant, ed., *Robert Boyle's Experiments in Pneumatics* (Cambridge: Harvard University Press, 1970), 18. Although Conant notes that "there is no such thing as *the* scientific method," his understanding of scientific work conforms to the hypothetico-deductive method of theory confirmation. (See Carl Hempel, "Studies in the Logic of Confirmation," in his *Aspects of Scientific Explanation* [New York: Free Press, 1965].) Conant defines science as a series of conceptual schemes or theories arrived at when broad working hypotheses become conceptual schemes. From a broad working hypothesis, he says, "we can deduce . . . many *consequences*, each of which can be tested by experiment." These consequences are limited working hypotheses which take the form "if, then." When combined with "tacit assumptions" including laws, these hypotheses can be tested by experiments. And Conant concludes that if "these tests confirm the deductions in a number of instances, the accumulating evidence tends to confirm the broad working hypothesis and the hypothesis may soon come to be accepted as a new conceptual scheme" (Conant, *Boyle's Experiments*, 4–7). Conant's reconstruction of the discovery of Boyle's Law conforms well to the hypothetico-deductive model of confirmation.

8. References are to Thomas Hobbes, *A Physical Dialogue of the Nature of the Air*, trans. Simon Schaffer, in Shapin and Schaffer, *Leviathan*, 354–355.

9. "One cannot help, however, but be somewhat skeptical of the high degree of accuracy reported by Perier. To be able to repeat the Torricellian experiment so that there was less than a twelfth of an inch (one 'line') difference in successive reading, as Perier claimed, is remarkable. The accidental intrusion of a slight amount of air is very difficult to avoid. The report of Perier's was written, it must be remembered, before standards of accurate reporting in science had been established. . . . It may be that Perier, persuaded of the reality of the large differences in height of the mercury column at the top and bottom of the mountain, succumbed to the temptation of making his argument appear convincing by recording exact reproducibility of his results on repeated trials" (Conant, *Boyle's Experiments*, 16–17).

10. Franciscus Linus, *Tractatus de corporum inseparabilitate*, 115–117. Quoted in Shapin and Schaffer, *Leviathan*, 159.

3. ECONOMICS, POLITICS, AND RELIGION: STUART CONFLICTS WITH PARLIAMENT

1. Christopher Hill, *The Century of Revolution 1603–1714* (London: Thomas Nelson and Sons, 1961), 31ff.
2. Godfrey Davies, *The Early Stuarts, 1603–1660* (Oxford: Clarendon Press, 1937), 172, and George Macaulay Trevelyan, *England under the Stuarts* (New York: G. P. Putnam's Sons, 1925), 187.

4. CIVIL WAR APPROACHES

1. Godfrey Davies, *The Early Stuarts, 1603–1660* (Oxford: Clarendon Press, 1937), 123.
2. *Memoirs of the Verney Family*, ed. Frances Parthenope Verney (1892), vol. 2, 69. *Memoirs of Edmund Ludlow*, ed. C. H. Firth (1894), vol. 1, 96. Both quoted in Davies, *The Early Stuarts*, 124–125.

5. THE INTERSECTION OF CLASS AND GENDER POLITICS

1. Christopher Hill, *Intellectual Origins of the English Revolution* (Oxford: Oxford University Press, 1982, originally published 1965), 40ff.
2. H. N. Brailsford, *The Levellers and the English Revolution*, ed. Christopher Hill (Nottingham: Spokesman Books, 1976), 314. Brailsford mentions only the occupations of Leveller men.
3. J. Dod and R. Cleaver, *A Godly Forme of Household Governement . . .* (London, 1614), sig. L5r-v. Quoted in Keith Thomas, "Women and the Civil War Sects," *Past and Present* 13 (1958), 43.
4. William Gouge, *Of Domesticall Duties* (Amsterdam and Norwood, N.J.: Walter J. Johnson, 1976; facsimile of the 1622 edition), 268.
5. Quoted in Thomas, "Women," 44. Brailsford notes that the radical gender politics of the Levellers mark sectarian influence: "[I]n their attitude to women the Levellers were ahead of their time. They encouraged women to play their part in politics side by side with their husbands and brothers, because they believed in the equality of all 'made in the image of God.' This was, indeed, an article of their religious creed, which reflected the influence of the Anabaptists among them. Everyone knows that however low the position of women sank round them, the Quakers always preached and practised equality. But few of us remember that they were following the example which their forerunners the Anabaptists had set from the early days of the sixteenth century onward. In their community women had an equal standing, an equal right to pray and speak at its meetings. So many of the Levellers were members of this sect that it must have seemed natural to practise on weekdays what they taught on Sundays" (Brailsford, *Levellers*, 317).
6. Elaine C. Huber, "'A Woman Must Not Speak': Quaker Women in the English Left Wing," in *Women of Spirit: Female Leadership in the Jewish and Christian*

Traditions, ed. Rosemary Ruether and Eleanor McLaughlin (New York: Simon and Schuster, 1977), 158.

7. Christopher Hill, *The Century of Revolution 1603–1714* (London: Thomas Nelson and Sons, 1961), 203.

8. Quoted in Patricia Higgins, "The Reactions of Women," in *Politics, Religion and the English Civil War,* ed. Brian Manning (New York: St. Martin's Press, 1973), 183.

6. THE BOYLE FAMILY'S RELIGIOUS AND CLASS POLITICS

1. Nicholas Canny, *The Upstart Earl: A Study of the Social and Mental World of Richard Boyle, First Earl of Cork, 1566–1643* (Cambridge, London, New York: Cambridge University Press, 1982), 36.

2. Quoted in Flora Masson, *Robert Boyle: A Biography* (London: Constable and Company, 1914), 130.

3. George Macaulay Trevelyan, *England under the Stuarts* (New York: G. P. Putnam's Sons, 1925), 276.

4. James R. Jacob, *Robert Boyle and the English Revolution: A Study in Social and Intellectual Change* (New York: Burt Franklin, 1977), 19.

7. MORE CLASS AND GENDER POLITICS

1. Godfrey Davies, *The Early Stuarts, 1603–1660* (Oxford: Clarendon Press, 1937), 145–146.

2. H. N. Brailsford, *The Levellers and the English Revolution,* ed. Christopher Hill (Nottingham: Spokesman Books, 1976), 245.

3. C. V. Wedgwood, *The Trial of Charles I* (Penguin: Harmondsworth, 1983), 29.

4. Quoted in Patricia Higgins, "The Reactions of Women," in *Politics, Religion and the English Civil War,* ed. Brian Manning (New York: St. Martin's Press, 1973), 201.

5. Lockyer was a common soldier executed for demanding some of his pay. He and other soldiers refused to obey orders until they "received some of their arrears" (Brailsford, *Levellers,* 506–507).

6. J. Howell, quoted in C. Hill and E. Dell, *The Good Old Cause* (London: Frank Cass, 1969), 278.

7. In Thomas Carlyle, *The Letters and Speeches of Oliver Cromwell,* ed. S. C. Lomas (London: Methuen, 1904), vol. 2, 342–343.

8. R. E. W. Maddison, *The Life of the Honourable Robert Boyle* (New York: Barnes and Noble, 1969), 98.

8. BOYLE'S GENDER POLITICS

1. In his autobiographical essay "An Account of Philaretus during His Minority," Boyle wrote, "When once Philaretus was able, without danger, to support the incommodities of a remove, his father, who had a perfect aversion for their fondness, who used to breed their children so nice and tenderly, that a hot sun, or a good shower of rain, as much endangers them, as if they were made of butter,

or of sugar, sends him away from home, and commits him to the care of a country nurse, who by early inuring him by slow degrees, to a coarse but cleanly diet, and to the usuall passions of the air, gave him so vigorous a complexion, that both hardships were made easy to him by custome, and the delights of conveniences and ease were endeared to him by their rarity." He goes on to mention the death of his mother (when he was about three years old), whom, he says, he never knew (*W*, I, xiii).

2. BP, xxxvii. 129–132.

3. Frances E. Dolan, "Taking the Pencil out of God's Hand: Art, Nature, and the Face-Painting Debate in Early Modern England," *PMLA* 108:2 (March 1993), 229, 225, 234.

4. An excellent discussion of Dury as well as of Hartlib and the third member of their group, Comenius, is found in H. R. Trevor-Roper, "Three Foreigners and the Philosophy of the English Revolution," *Encounter* 14 (1960), 3–20.

5. James R. Jacob, *Robert Boyle and the English Revolution: A Study in Social and Intellectual Change* (New York: Burt Franklin, 1977), 93.

6. *W*, V, 285. That "*Theodora* was not published until 1687" is "further evidence for interpreting it as politico-religious allegory because it must then have been intended and read as an attack on the Catholic policies of James II. More evidence for the allegorical significance of the work is the following undated remark found among Boyle's papers: 'Whatever Papers be found of mine relateing to Theodora, I desire may be burnt without fail.' Such probably were Boyle's fears in 1648–49" (Jacob, *Robert Boyle*, 93).

9. BOYLE'S BACKGROUND READING

1. R. E. W. Maddison, *The Life of the Honourable Robert Boyle* (New York: Barnes and Noble, 1969), 66 and 66n.

2. Marie Boas, *Robert Boyle and Seventeenth-Century Chemistry* (Cambridge: Cambridge University Press, 1958), 21.

3. Frances A. Yates, *Giordano Bruno and the Hermetic Tradition* (New York: Vintage Books, 1964), 13, 2, 4.

4. Quoted in Peter Rattansi, "The Social Interpretation of Science in the Seventeenth Century," in *Science and Society*, ed. Peter Mathias (Cambridge: Cambridge University Press, 1972), 9–10.

5. Quoted in D. P. Walker, *Spiritual and Demonic Magic* (Notre Dame, Ind.: University of Notre Dame Press, 1975), 13.

6. Walter Pagel, *Paracelsus* (Basel: Karger Press, 1982), 84, 85.

7. Walter Pagel, *Joan Baptista Van Helmont* (Cambridge, London, New York: Cambridge University Press, 1982), 32, 97.

10. BOYLE'S HERMETICISM, MAGIC, AND ACTIVE PRINCIPLES

1. P. M. Rattansi, "The Intellectual Origins of the Royal Society," *Notes and Records of the Royal Society* 23 (1968), 131.

2. Harold Fisch, "The Scientist as Priest: A Note on Robert Boyle's Natural

Theology," *Isis* 44:137 (1953), 254. Fisch points to striking similarities among Boyle's texts, those of Thomas Browne's Hermetic work, *Religio Medici,* and that of the "thorough-going alchemist," Elias Ashmole.

3. Michael Hunter, "Alchemy, Magic and Moralism in the Thought of Robert Boyle," *British Journal for the History of Science* 23:79 (1990), 387–410. The five pages comprise British Library Add. MS 4229, fols. 60–63, endorsed "Mr Boyle's papers dictated by him, copied by Dr Burnet, relating to his Life."

4. Hunter, "Alchemy," 390–391, 396–398, quoting Elias Ashmole, *Theatrum Chemicum Britannicum* (London, 1652), Sig. Blv.

5. A list of these is found in Royal Society ms. 198, fol. 143v.

6. Marie Boas, *Robert Boyle and Seventeenth-Century Chemistry* (Cambridge: Cambridge University Press, 1958), 102; Hunter, "Alchemy," 404.

7. Marie Boas, "An Early Version of Boyle's *Sceptical Chemist,*" *Isis* 45:140 (July 1954), 154.

8. John Henry, "Occult Qualities and the Experimental Philosophy: Active Principles in Pre-Newtonian Matter Theory," *History of Science* 24:66 (1986), 352; the argument is pursued in his "Boyle and Cosmical Qualities," in *Robert Boyle Reconsidered,* ed. Michael Hunter (Cambridge: Cambridge University Press, 1994), 119–138.

9. Henry, "Occult Qualities," 356. Henry More, *Enthusiasmus triumphatus: Or, a brief discourse of the nature, causes, kinds, and cure of enthusiasm* (London, 1656), 31. Henry argues that Boyle's voluntaristic theology led him to accept active matter. But see Antonio Clericuzio, "A Redefinition of Boyle's Chemistry and Corpuscular Philosophy," *Annals of Science* 47:6 (1990), 573. Clericuzio argues that "Boyle's voluntaristic theology, which Henry has invoked to support his thesis, can in fact be used to prove precisely the opposite." The world depends on God's absolute power, so there was no natural necessity for the continuation of motion (or for the conservation of matter).

11. HERMETICISM, HYLOZOISM, AND RADICAL POLITICS

1. P. M. Rattansi, "Paracelsus and the Puritan Revolution," *Ambix* 11:1 (1963), 24–32; and P. M. Rattansi, "The Intellectual Origins of the Royal Society," *Notes and Records of the Royal Society* 23 (1968), 129–143. See also his "The Helmontian-Galenist Controversy in Restoration England," *Ambix* 12:1 (1964), 1–23; "The Social Interpretation of Science in the Seventeenth Century," in *Science and Society,* ed. Peter Mathias (Cambridge: Cambridge University Press, 1972), 1–32; and "Recovering the Paracelsian Milieu," in *Revolutions in Science: Their Meaning and Relevance,* ed. William R. Shea (Canton, Mass.: Science History Publications/U.S.A., 1988), 1–26. See also James R. Jacob and Margaret C. Jacob, "The Anglican Origins of Modern Science: The Metaphysical Foundations of the Whig Constitution," *Isis* 71:257 (1980), 251–267; and the following works by James R. Jacob: "The Ideological Origins of Robert Boyle's Natural Philosophy," *Journal of European Studies* 2:1 (1972), 1–21; "Robert Boyle and Subversive Religion in the Early Restoration," *Albion* 6:4 (1974), 175–93; *Robert Boyle and the English Revolution: A Study in*

Social and Intellectual Change (New York: Burt Franklin, 1977); "Boyle's Circle in the Protectorate: Revelation, Politics, and the Millennium," *Journal of the History of Ideas* 38:1 (1977), 131–140; and "Boyle's Atomism and the Restoration Assault on Pagan Naturalism," *Social Studies of Science* 8:2 (1978), 211–233. See also Elizabeth Potter, "Modeling the Gender Politics in Science," in *Feminism and Science*, ed. Nancy Tuana (Bloomington and Indianapolis: Indiana University Press, 1989), 132–146, and "Gender and Epistemic Negotiation," in *Feminist Epistemologies*, ed. L. Alcoff and E. Potter (Routledge: New York, 1993), 161–182.

2. John Henry, "Occult Qualities and the Experimental Philosophy: Active Principles in Pre-Newtonian Matter Theory," *History of Science* 24:66 (1986), 365n.

3. Antonio Clericuzio, "A Redefinition of Boyle's Chemistry and Corpuscular Philosophy," *Annals of Science* 47:6 (1990), 566.

4. Michael Hunter, "Alchemy, Magic, and Moralism in the Thought of Robert Boyle," *British Journal for the History of Science* 23 (1990), 407; BP 25, 285; 38, fol. 160; 19, fols. 187v–188.

5. Thomasio Campanella, *On the Sense and Feeling in All Things and on Magic*, in *Renaissance Philosophy*, vol. 1, ed. and trans. Arturo B. Fallico and Herman Shapiro (New York: The Modern Library, 1967), 362.

6. Quoted by Christopher Hill in *The World Turned Upside Down* (New York: Penguin Books, 1975), 142.

7. *A Ternary of Paradoxes* (London, 1650), *Prolegomena*. Quoted in Rattansi, "Paracelsus," 26.

8. See his *Physiologia Epicuro-Gassendo-Charltoniana* (London, 1654), 58.

12. BOYLE'S CONCERN OVER THE SECTARIES

1. This quotation is from Boyle's *Occasional Reflections*. Jacob notes that this work was "written some time between the regicide and the restoration" (James R. Jacob, "The Ideological Origins of Robert Boyle's Natural Philosophy," *Journal of European Studies* 2:1 [1972], 10n) and, indeed, we find in the "advertisement" prefacing it the remark that "A reader, that is not unattentive, may easily collect from what he will meet with in some of the ensuing discourses, that they were written several years ago, under an usurping government, that then prevailed" (*W*, II, 390).

2. Christopher Hill, *The World Turned Upside Down* (New York: Penguin Books, 1975), 185.

3. James R. Jacob, "Boyle's Circle in the Protectorate: Revelation, Politics, and the Millennium," *Journal of the History of Ideas* 38:1 (1977), 134; *W*, II, 276.

4. According to the sermon notes, the Fifth Monarchists "considered Cromwell a usurper, an offender against God, who in return would rain down destruction upon him and his state. This for the Fifth Monarchists would be tantamount to the apocalypse and the advent of the millennium in which they, God's elect, would rule" (Jacob, "Boyle's Circle," 137–138). Jacob argues that the sermons at which the notes were taken were part of the Fifth Monarchists' efforts to join with other groups to overthrow the Protectorate. This joint attempt at overthrow never occurred, but the Fifth Monarchists did revolt in 1661 after the restoration of the

monarchy. For a detailed exposition of the beliefs and activities of the Fifth Monarchists, see P. G. Rogers, *The Fifth Monarchy Men* (London: Oxford University
Press, 1966).

5. See *W*, I, xxxii–xxxiii, where Boyle says of sectarian opinions that "some
[are] digged out of those graves where the condemning decrees of primitive councils had long since buried them; others newly fashioned in the forge of their own
brains; but the most being new editions of old errors."

6. See Steven Shapin and Simon Schaffer, *Leviathan and the Air-Pump: Hobbes,
Boyle, and the Experimental Life* (Princeton: Princeton University Press, 1985),
207ff., for a discussion of Boyle's insistence against Henry More that experimental
philosophy is a separate enterprise from theology and should not be used to make
theological points. More had used some of Boyle's experiments to prove the existence of a Spirit of Nature. Of course, Boyle himself used his experiments to argue
against the personification of Nature and the animation of matter (for example,
W, II, 36–38). My own view is that Boyle was inconsistent in his enforcement of
the boundary between theology and science because he was quite consistent in
rejecting the personification of Nature. He selectively enforced the boundary on
behalf of his own religious and associated political views and against religious and
associated political views that he found objectionable.

7. Henry More, *Enthusiasmus Triumphatus: Or, a Brief Discourse of the Nature,
Causes, Kinds, and Cure of Enthusiasm* (London, 1656), 3; quoted in John Henry,
"Occult Qualities and the Experimental Philosophy: Active Principles in Pre-
Newtonian Matter Theory," *History of Science* 24:66 (1986), 356.

8. *Calendar of State Papers, Ireland, 1663–1665*, 100–101, 662, and 669. Cited
in James R. Jacob, "Robert Boyle and Subversive Religion in the Early Restoration," *Albion* 6:4 (1974), 279.

9. *A Free and Impartial Censure of the Platonick Philosophie* (Oxford, 1666), 75–
76. Quoted in Jacob, "Subversive Religion," 279. Jacob suggests that Boyle also
had this group in mind when he attacked a "sect of men" who have "lately sprung
up," since Heydon's circle answers the description Boyle gives of the contemporaries he wishes to refute.

10. More, *Enthusiasmus Triumphatus*, 1; quoted in Henry, "Occult Qualities,"
356. Apparently Isaac Newton agreed. According to the Jacobs, Newton repudi-
ated both the Cartesian and "vulgar" or Aristotelian physical accounts of bodies
because they both lead to atheism: "Newton repudiates Descartes's definition of
body as extension because it does 'manifestly offer a path to Atheism'; likewise
he repudiates 'the vulgar notion (or rather lack of it) of body . . . in which all
the qualities of the bodies are inherent' because it too leads directly to atheism."
The Jacobs argue that Newton, like Boyle, defined his metaphysics in response to
the threat of atheism posed by vitalistic materialism as represented by Henry
Stubbes ("that most infamous Restoration heretic and subversive"), by the materialism of Hobbes, and by the dualism of Descartes (James R. Jacob and Margaret C.
Jacob, "The Anglican Origins of Modern Science: The Metaphysical Foundations
of the Whig Constitution," *Isis* 71:257 [1980], 262). Quotations are from Newton's
"De Gravitatione et aequipondo fluidorum."

11. Michael Hunter, "Science and Heterodoxy: An Early Modern Problem Reconsidered," in *Reappraisals of the Scientific Revolution*, ed. David C. Lindberg and Robert S. Westman (Cambridge: Cambridge University Press, 1990), 445.

13. BOYLE'S OBJECTIONS TO HYLOZOISM

1. Richard S. Westfall suggests that the first part of *Some Considerations* was written between 1649 and 1654, but it is not possible to tell how much of this work was modified for publication in 1663. See his "Unpublished Boyle Papers Relating to Scientific Method.—I," *Annals of Science* 12:1 (1956), 65.

2. In *Some Considerations*, Boyle also carefully distinguishes himself from Helmont and Paracelsus by stressing the importance of reason in studying nature, rather than relying on revelation of it in dreams and visions. Although God can reveal himself to people in supernatural ways, such supernatural gifts relate to another world (not to this one, as the sectaries claim). Thus, when we study nature, we should consult its author, Boyle says, "from whom descends every good and every perfect gift; not only those supernatural graces, that relate to another world, but those intellectual endowments, that qualify men for the prosperous contemplation of this. . . . And though I dare not affirm, with some of the Helmontians and Paracelsians, that God discloses to men the great mystery of chymistry by good angels, or by nocturnal visions . . . yet persuaded I am, that the favour of God does . . . vouchsafe to promote some men's proficiency in the study of nature . . . principally, by directing them to those happy and pregnant hints, which an ordinary skill and industry may so improve, as to do such things, and make such discoveries by virtue of them, as both others, and the person himself . . . would scarce have imagined to be possible" (*W*, II, 61).

3. Rose-Mary Sargent, *The Diffident Naturalist: Robert Boyle and the Philosophy of Experiment* (Chicago: University of Chicago Press, 1995), 109, 87n.

14. EXPERIMENTAL SUPPORT FOR THE CORPUSCULAR PHILOSOPHY

1. Quoted in Steven Shapin and Simon Schaffer, *Leviathan and the Air-Pump: Hobbes, Boyle, and the Experimental Life* (Princeton: Princeton University Press, 1985), 160; the words in brackets are my additions.

2. See the excellent discussion of Boyle's work on cohering marbles in Shapin and Schaffer, *Leviathan*, 187–201. They date Boyle's new experiment to 1662; see 194n. And they point out that a "critic with a mind to make trouble could have objected that the attached weights served to ensure the marbles' separation, and not merely to correct for the forces working against a 'natural' outcome" (195).

3. Without an explanation, Hobbes argued, the very notion of the air's spring was inconceivable or absurd, since it entailed that atoms spontaneously move themselves. That is, Boyle analogized the corpuscles of air to wool fleece: fleece can be compressed, for example, by having more fleece piled upon it. But when the upper fleece is removed, the wool springs back. An air particle is like a hair of wool or like the steel plate of a crossbow that returns to its original shape when released (*W*, I, 11–12). But this, said Hobbes, explains nothing: "Both these fan-

tasies, the gravity of the air as well as the elastic force or spring [*antitupia*] of the air, were dreams. For if spring were allowed by them to be something in the threads of the air, and they were to search for something by which, when somewhat curved yet at rest, the threads would be moved again to straightness: if they wish to be taken for physicists, they would have to assign some possible cause for it" (quoted in Shapin and Schaffer, *Leviathan*, 377).

Failing a mechanical explanation, Boyle must either admit that the notion of the air's spring is absurd or admit that something can move itself. "For you suppose," Hobbes charges, "that the air particle, which certainly stays still when pressed, is moved to its own restitution, assigning no cause for such a motion, except that particle itself" (quoted in Shapin and Schaffer, *Leviathan*, 363). This supposition, Shapin and Schaffer point out, would put Boyle in the same camp with the sectaries whose "'vulgar' and dangerous conception of self-moving matter" he wished to undermine (141). We should note that Boyle in turn tried to attach the label "enthusiast" to Hobbes by charging that the earthy particles postulated in his physics enjoyed innate circular motion; in *An Examen of Mr. T. Hobbes* he says, "I shall have occasion to shew anon, that Mr. Hobbes himself, whatever he says in this place, does elsewhere ascribe a motion of their own to multitudes of terrestrial corpuscles" (*W*, I, 195).

16. THE PRODUCTION OF AN ALTERNATIVE LAW

1. Conor Reilly, *Francis Line S.J.: An Exiled English Scientist (1595–1675)* (Rome: Institutum Historicum, S.I., 1969), 83, 122. Reilly tells us that Linus's hypothesis was taken up by the Cartesian Gilbert Clerke, who agreed that "some sort of *funiculus* could exist, but this was, he said, really Descartes' aether, and not Line's 'subtle body.' Clerke's *funiculus* was made up of the 'purest parts of the air . . . which have been drawn into the top of the barometer tube through the pores of the glass'" (87). And Lord Chief Justice Matthew Hale, like Linus, "accepted the rules of the experimental game," but adopted plenism and Linus's funicular hypothesis against Boyle's hypothesis of the air's spring (Steven Shapin and Simon Schaffer, *Leviathan and the Air-Pump: Hobbes, Boyle, and the Experimental Life* [Princeton: Princeton University Press, 1985], 50–51).

17. METHODOLOGICAL CONSIDERATIONS

1. BP, XXXV; quoted in Richard Westfall, "Unpublished Boyle Papers Relating to Scientific Method.—II," *Annals of Science* 12:2 (1956), 116–117.

2. Laurens Laudan, "The Clock Metaphor and Probabilism: The Impact of Descartes on English Methodological Thought, 1650–65," *Annals of Science* 22:2 (1966), 87.

3. Larry Laudan, *Science and Values* (Berkeley and Los Angeles: University of California Press, 1984), 28.

4. Elizabeth Potter, "Underdetermination Undeterred," in *Feminism, Science, and the Philosophy of Science*, ed. Lynn Hankinson Nelson and Jack Nelson (London: Kluwer Academic Publishers, 1996), 121–138.

18. "THE DATA ALONE PROVED BOYLE'S HYPOTHESIS"

1. Hans Reichenbach, *Experience and Prediction* (Chicago: University of Chicago Press, 1938), especially 3–7, where Reichenbach introduces the distinction in connection with the notion of rational reconstruction.

2. Rose-Mary Sargent, *The Diffident Naturalist: Robert Boyle and the Philosophy of Experiment* (Chicago: University of Chicago Press, 1995), 16, 55, and 56.

3. Laurens Laudan, "The Clock Metaphor and Probabilism: The Impact of Descartes on English Methodological Thought, 1650–65," *Annals of Science* 22:2 (1966), 81–82.

4. Carl G. Hempel, "Turns in the Evolution of the Problem of Induction," *Synthese* 43:3 (1981), 393.

5. Richard Rudner, "The Scientist qua Scientist Makes Value Judgments," *Philosophy of Science* 20:1 (1953), 1–6. Richard Jeffrey attempted to defend the value neutrality of the scientist by arguing in response to Rudner that the scientist qua scientist does not accept or reject hypotheses, he merely calculates their probabilities! Hempel rejects this "startling" attempt and notes that even Jeffrey had no response to Rudner's argument that, in calculating the probability of a hypothesis, the scientist must accept hypotheses describing the evidence he uses in his calculations, and that this acceptance in turn depends upon how important it is not to make a mistake. See Richard C. Jeffrey, "Valuation and Acceptance of Scientific Hypotheses," *Philosophy of Science* 22:3 (1956), 237–246.

6. Sargent does recognize that "Boyle's concurrence of probabilities is a decidedly vague epistemological notion" but she recognizes this only with regard to determining when the evidence is relevant (208). I suggest that it is especially vague on the most important point, viz., when the evidence proves the hypothesis.

7. Steven Shapin and Simon Schaffer, *Leviathan and the Air-Pump: Hobbes, Boyle, and the Experimental Life* (Princeton: Princeton University Press, 1985), 50.

8. "[W]e have attempted to show (1) that the solution to the problem of knowledge is political; it is predicated upon laying down rules and conventions of relations between men in the intellectual polity; (2) that the knowledge thus produced and authenticated becomes an element in political action in the wider polity . . . ; (3) that the contest among alternative forms of life and their characteristic forms of intellectual product depends upon the political success of the various candidates in insinuating themselves into the activities of other institutions and other interest groups. He who has the most, and the most powerful, allies wins" (Shapin and Schaffer, *Leviathan*, 342).

9. Steven Shapin, "History of Science and Its Sociological Reconstruction," *History of Science* 20:49 (1982), 181.

10. See, for example, Thomas S. Kuhn, *The Structure of Scientific Revolutions* (Chicago: University of Chicago Press, 1962, 1970), 185–186; Hempel, "Turns in the Evolution," 399ff.

11. See P. M. Rattansi, "Paracelsus and the Puritan Revolution," *Ambix* 11:1

(1963), 24–32; P. M. Rattansi, "The Intellectual Origins of the Royal Society," *Notes and Records of the Royal Society* 23 (1968); Peter Rattansi, "The Social Interpretation of Science in the Seventeenth Century," in *Science and Society,* ed. Peter Mathias (Cambridge: Cambridge University Press, 1972), 1–32; John Henry, "Occult Qualities and the Experimental Philosophy: Active Principles in Pre-Newtonian Matter Theory," *History of Science* 24:66 (1986), especially 365n; Shapin and Schaffer, *Leviathan,* especially chapters 3, 5, 7, and 8; James R. Jacob, "Boyle's Circle in the Protectorate: Revelation, Politics, and the Millennium," *Journal of the History of Ideas* 38:1 (1977), 131–140; James R. Jacob, *Robert Boyle and the English Revolution: A Study in Social and Intellectual Change* (New York: Burt Franklin, 1977); and James R. Jacob and Margaret C. Jacob, "The Anglican Origins of Modern Science: The Metaphysical Foundations of the Whig Constitution," *Isis* 71:257 (1980).

12. Lynn Nelson, *Who Knows: From Quine to a Feminist Empiricism* (Philadelphia: Temple University Press, 1990), 187.

13. See Kathryn Pyne Addelson, "The Man of Professional Wisdom," in *Discovering Reality: Feminist Perspectives on Epistemology, Metaphysics, Methodology, and Philosophy of Science,* ed. Sandra Harding and Merill B. Hintikka (Dordrecht, Boston, and London: D. Reidel, 1983), 182; Anne Fausto-Sterling, *Myths of Gender: Biological Theories about Women and Men* (New York: Basic Books, 1985), 11–12 and passim, but esp. chapter 7; Bonnie B. Spanier, *Impartial Science: Gender Ideology in Molecular Biology* (Bloomington and Indianapolis: Indiana University Press, 1995); Helen Longino, *Science as Social Knowledge* (Princeton, N.J.: Princeton University Press, 1990), passim, but esp. chapters 5–8.

19. GOOD SCIENCE

1. We will use the terms "constitutive" and "contextual" values, following the nomenclature of Helen Longino in her *Science as Social Knowledge* (Princeton, N.J.: Princeton University Press, 1990).

2. Elizabeth Potter, "Modeling the Gender Politics in Science," in *Feminism and Science,* ed. Nancy Tuana (Bloomington and Indianapolis: Indiana University Press, 1989), 132–146; Elizabeth Potter, "Gender and Epistemic Negotiation," in *Feminist Epistemologies,* ed. L. Alcoff and E. Potter (Routledge: New York, 1993), 161–182.

3. Mary B. Hesse, *The Structure of Scientific Inference* (Berkeley and Los Angeles: University of California Press, 1974). For an exposition of Hesse's model which is accessible to non-philosophers, see David Bloor, "Durkheim and Mauss Revisited: Classification and the Sociology of Knowledge," *Studies in the History and Philosophy of Science* 13 (1982), 267–297.

4. David Bloor argues that coherence conditions should include any social interests that influence the choice a scientist makes between competing hypotheses (Bloor, "Durkheim and Mauss Revisited"). Although he does not mention gender considerations, he might include androcentric or sexist social interests as coherence conditions. However, there is a problem with the notion of "social interest" here: a woman scientist might well adopt a sexist or androcentric coherence condition even when it is against her social interest to do so. See also Longino, *Science,*

esp. chapter 3. On the model set out here, what she refers to as "background assumptions" determine scientists' classificatory decisions.

5. The sociobiological theories of Sara Hrdy, Nancy Tanner, and Adrienne Zihlman beautifully illustrate this point. See Sara Blaffer Hrdy, *The Woman That Never Evolved* (Cambridge, Mass.: Harvard University Press, 1981); Nancy Tanner, *On Becoming Human* (Cambridge: Cambridge University Press, 1981); Nancy Tanner and Adrienne Zihlman, "Women in Evolution, Part I: Innovation and Selection in Human Origins," *Signs* 1:3 (1976), 585–608; Adrienne Zihlman, "Women in Evolution, Part II: Subsistence and Social Organization among Early Hominids," *Signs* 4:1 (1978), 4–20.

6. Both the model-theoretic theory of theories of Bas Van Fraassen and Ronald Giere's decision-theoretic model of science could be helpful here. See Bas C. Van Fraassen, *The Scientific Image* (Oxford: Clarendon Press, 1980), and Ronald N. Giere, *Explaining Science* (Chicago and London: University of Chicago Press, 1988).

7. See, for example, studies listed in Steven Shapin, "History of Science and Its Sociological Reconstruction," *History of Science* 20:49 (1982), and many studies found in the pages of the journal *Social Studies of Science*.

Index

with children, 78; and relationships with men, 3–9, 13; in the Renaissance, 5; in the Royal Society, 16; as scientists, 16–19, 20–21; sectarian, 54; and transvestism, 14; upper-class, 77–79; violent demonstrations by, 59–

60; as wives, 13, 54, 74–75. *See also* Feminists; Gender; Men
World Soul, 97
World Spirit, 106, 107–108, 156; Robert Boyle on, 120, 124–126. *See also* Animism

*   *   *

ELIZABETH POTTER is the Alice Andrews Quigley Professor of Women's Studies at Mills College. She is co-editor of *Feminist Epistemologies* and author of numerous articles in feminist epistemology and feminist philosophy of science.